DATE			

BETWEEN GOD

and

GANGSTA RAP

Also by Michael Eric Dyson

Reflecting Black: African-American Cultural Criticism

Making Malcolm: The Myth and Meaning of Malcolm X

BETWEEN GOD and GANGSTA RAP

bearing witness to black culture

Michael Eric Dyson

New York Oxford / **OXFORD UNIVERSITY PRESS** / 1996

Oxford University Press

Oxford   New York
Athens   Auckland   Bangkok   Bombay
Calcutta   Cape Town   Dar es Salaam   Delhi
Florence   Hong Kong   Istanbul   Karachi
Kuala Lumpur   Madras   Madrid   Melbourne
Mexico City   Nairobi   Paris   Singapore
Taipei   Tokyo   Toronto

and associated companies in
Berlin   Ibadan

Copyright © 1996 by Michael Eric Dyson

Published by Oxford University Press, Inc.,
198 Madison Avenue, New York, New York 10016

Oxford is a registered trademark of Oxford University Press

Library of Congress Cataloging–in–Publication Data
Dyson, Michael Eric.
Between God and gangsta rap : bearing witness to Black culture /
Michael Eric Dyson.
p.   cm.   Includes index.
ISBN   0–19–509898–6
1. Afro-Americans—Social conditions—1975-
2. Afro-American arts.   I. Title.
E185.86.D94   1996   305.896'073—dc20   95-20779

1 3 5 7 9 8 6 4 2

Printed in the United States of America
on acid-free paper

For our children

Maisha

Michael

and

Mwata

and

To the loving memory of my "Son"

Stephen Dale Boliver

(1964-1994)

A Baptist Beat

A mixed congregation: sinners, worshippers,
Hustlers, survivors. All that terrible energy,
Locked in, trying to blend. Such a gathering
Of tribes has little, if any, use for a silk-robed choir.
Members bring their own noise, own souls.
Any Avenue Crew will tell you; nothing comes closer
To salvation than this. Here, there is not talk of Judgment,
No fear. Every now & then, an uninformed God
Will walk in, bear witness, mistake kangol
For halo, all those names for unwanted bodies
Being called home, arms raised to testify, waving
From side to side, fists flying like bullets, bullets
Like fists. Above the snare: two sticks make the sign
Of the cross then break — a divorced crucifix.
The tambourine shakes like a collection plate.
This pastor wants to know who's in the house,
Where we're from, are we tired yet, ready to quit?
We run down front, scream & shout, "Hell no,
We ain't ready to go!" The organ hesitates,
Fills the house with grace, good news, resurrection
& parole, a gospel of chords rising like souls.
Up, up, up up, down, down. Up, up, up up,
Down. Up, up, up up, down, down.
Up, up, up up, down. The cowbell's religious beat,
A prayer angel ushered through the dangerous air.

Thomas Sayers Ellis

PREFACE

Speaking in Tongues

"Anyone who speaks in tongues should pray for the ability to interpret."   **1 Corinthians 14: 13**

During the last decade, blacks have tried to shatter a chain of challenges to our communal flourishing from within and outside of black life. The chain is composed of many links that are both familiar and frightening. It includes the conservative assault on affirmative action. The fateful lapse in black political and civil rights leadership. The resurgence of racism and xenophobia. The development of underground drug economies and gang violence in the inner city. The painful collapse of a liberal consensus on race. The increase in sexual violence among teens. The chilling rise in rates of black imprisonment. The expanding material misery of the working-poor and ghetto-poor. And much, much more.

The links in that chain are durable and forged by a combination of societal constraint and personal choice. Those links have barely been weakened by our strenuous efforts to change the cultural habits that weigh us down or to criticize the society that hampers our progress. There's no use for me to feign neutrality on the question, however, of which comes first: the chicken of structural factors that shape social life and individual behavior or the egg of human responsibility to alter our conditions? In my mind the chicken comes first.

That doesn't mean that I'm not concerned about black folk being responsible for their lives and culture. I'm all for getting rid of the dangerous and destructive self-pity and victim-addiction that unquestionably plague some black people. Such a posture certainly contradicts the way most blacks have zealously pursued self-improvement and societal enhancement. But two notes must be quickly added. First, the number of blacks who are unwilling to get off of their psychic duffs and strive against the odds to create a healthy life is greatly exaggerated. Most black folk work hard and observe deeply entrenched cultural habits of self-help. The real wonder is that more black people haven't given in to spiritual or psychological inertia.

Second, the social conditions that victimize black folk continue to exist; many black folk have simply found a way around them. Many of the ones who haven't should not be blamed for lack of ingenuity or will. Their failure may simply have more to do with the nature of the obstacles they faced than with their coping skills. Black folk should not be penalized for calling attention to the persistence of these obstacles. But the new national habit of demonizing *minority* victims has generated hostility toward those who make legitimate claims of personal or social suffering.

Note, too, that the social circumference for defining legitimate "victims" has considerably tightened. A pernicious consequence of this state of affairs is that our society is let off the hook for practices that inflict harm on blacks and other minorities without their opportunity to say so. If they speak up, they're labeled whining victims. If they remain quiet, they give tacit approval to their status, thus reinforcing their suffering. It's a vicious catch-22.

Troubling, too, are the signs of radical fissure within black cultures. Class tensions continue to brew between middle income and poor blacks. As more blacks become upwardly mobile and fan out into suburbia, the pattern of social life among blacks has dramatically changed. The geography of black identity and culture—how black folk are physically located and related to one another—has been greatly transformed by such patterns. Plus, ghetto communities have been intensely assaulted by the postindustrial collapse of segments of the American economy where blacks were once heavily employed. These features have driven wedges into black life at odd angles. The confusion and conflicts they have engendered—seen in the snobbishness of some middle-class blacks toward less well-off blacks, and in the resentment by some working-class members of middle-class flight—are increasing daily.

The sharp emergence of gender conflict in black life is equally troubling. The Anita Hill–Clarence Thomas debacle focused attention on the psychic and cultural injuries that black women have suffered in silence for centuries. The Mike Tyson rape case revealed just how confused some black men and women can be in automatically defending ancient beliefs about the "black-woman-as-seductress." In this case, Desiree Washington, the young woman Tyson was convicted of raping, fit the bill. When many black folk publicly prayed for Tyson and not Washington, indeed almost *against* Washington, it signaled the rejection of a liberating black religious ethic. Washington's treatment also symbolizes the low premium that black culture, and the rest of our society, often place on black women's physical beings. For many, Tyson's hulking proportion loomed larger than Washington's female frame in the cultural contest of valued bodies.

The recycling of tired debates about racial and cultural authenticity abound. These debates have taken many forms in many different forums, but they all come down to the same question: how can we define the Real Black Person? That question has bitterly divided black folk along a number of lines. It's viewed in the "unreal" bourgeois black versus the "real" ghetto black. It's glimpsed in the "unreal" integrationist black versus the "real" nationalist black. It's seen in the "unreal" gay or lesbian black versus the "real" heterosexual black. It's captured in the "unreal" multicultural black versus the "real" Afrocentric black. And so on.

The problem with most debates about authenticity, of course, is that they ignore the bewildering variety of expressions that characterize contemporary black culture. To be sure, the quest for the Real Black Person was initially set in motion by black folk defending themselves against the rigid rule of white stereotype and the suffocating effect of white prejudice. Black culture was defined by white folk in narrow ways. Black intellectuals began to defend black culture by defining its unique characteristics and edifying features. From that noble and necessary enterprise has evolved a sometimes mean-spirited effort to justify a questionable notion of racial authenticity, often by those who claim a special access to knowing just what that is. It is often a self-serving and wasteful indulgence.

Many of the divisions in black life—especially those based on gender, class, sexuality, authenticity, and generation—come together in debates about the virtues and vices of hip-hop culture. Hip-hop artists furiously debate the politics of authenticity; many artists have as a motto to "keep it real." That reality often has as much to do with the class and cultural identification of their music as it does with any notion of race. Though many of its artists have widely varying views on gender, hip-hop culture is notoriously sexist, often misogynistic, and at times willfully patriarchal. And hip-hop's homophobia is vicious and downright depressing.

There is, too, in debates about hip-hop a deepening aesthetic and rhetorical rift between older and younger blacks. How the music sounds and the speech it employs to make its points are often alienating to a more mature generation. This gulf is further revealed in the stylistic conflicts and ideological agendas that set the hip-hop generation apart from its parents. In particular, the controversies generated by gangsta rap—calling women bitches, celebrating violence, glamorizing sexual outlaw behavior—have reinforced *and* undermined sharp lines of division in black culture.

These divisions are not simply drawn between the old and young, but between the rich and the poor, between male and female, and between the schooled and the unlettered. The pattern of response to gangsta rap, however, is often unpredictable. For instance, there are well-to-do blacks who embrace the genre's unsparing social criticism, while there are working-class blacks who despise its message of violence and thuggery. The same complex response to the music applies as well to many who might appear either sympathetic or hostile because of their age, gender, education, and sexuality.

Despite its many problems, however, gangsta rap is not the most pressing problem that black folk face. The material suffering of the ghetto poor, the predicament of young black teens, the conservative onslaught against black interests and culture, and the devastating consequences of drug economies, for instance, are far more urgent. That is not to deny that gangsta rap's repulsive misogyny, virulent homophobia, and verbal violence are not cause for concern. They are.

But the vilification that gangsta rap has endured, and its mostly young black male artists right along with it, is far out of proportion to the problems it presents. The demonization of gangsta rappers is often a convenient excuse for cultural and political elites to pounce on a group of artists who are easy prey. The much more difficult task is to find out what conditions cause their anger and hostility. A more ennobling pursuit is to establish open lines of communication with gangsta rappers, to ask them why they speak and act as they do. A far more just response to their visions of life includes a sharp, substantive but sympathetic criticism of their work that weeds out the good from the bad. For black folk who have too often been dismissed, stigmatized, or silenced without a hearing, we should be wary of repeating such rituals of repression on our own kids.

As an ordained Baptist minister, university professor, radical democrat, and public intellectual, I daily encounter analyses and criticisms of black life by those outside and within black culture. As a result of the multiple kinship groups to which I belong, I live at the intersection of the varied expressions of American society and black culture. I believe that all segments of black cultures can learn from the others, and from the larger world. I believe that our best future will only be realized if we learn to listen and to speak to one another, and to those outside our culture. I have learned that in order to survive—a habit I have learned from wise black folk—I must be able to adapt and transform the languages I speak for the varied audiences I address. I have learned to speak in many tongues.

As a former pastor who still preaches frequently across the nation, I am nurtured by the songs, stories, and souls of black folk who contend that "this world ain't no friend to grace." I am reminded by black folk of faith that spirituality is a precious gift in a world of small vision and little hope. I have attended thousands of church services, prayer meetings, and religious conventions where the witness of ordinary people about a God "who can make a way out of no way" has brought light to the dark crevices of my mind and soul. I have listened for most of my life to brilliant black bards whose preaching fuses the art of the poet and prophet, the politician and the priest. And most of them have not advised black folk to retreat from the difficult work of living on earth even as they strive for heaven.

The language of faith is crucial because it affords human

beings the privilege of intimacy with the ultimate. The language of faith grounds human life in a set of religious narratives that transmit the sheer beauty and integrity of human existence, that affirm our birthright as children of God. Such language may be off-putting to secular citizens, but the best of religious traditions respects the humanity of all people, regardless of accidental features of their existence such as race, religion, class, gender, sexuality, or age. The language of faith is important to me because it guards against the corrosive cynicism and skepticism that might eat away at me during moments of profound depression about the way the world is going or when in doubt about how I fit into the larger scheme of things. The language of faith simply helps me bear witness to the God I believe in and to the people I love.

As a university professor, the language of the academy is extremely important to me. Although the university has come under attack for its practiced irrelevance to the larger society, and its intrinsic elitism, it is a wonderful place to be in the world. For too long, black folk have fought the perception that we weren't interested in literacy or learning, that we were little higher, intellectually, than the animals. But we struggled against such pernicious perceptions to produce a body of work that is important not only for black folk, but for the world. The belated acknowledgment of our talents can only encourage the further development of our intellectual and scholarly traditions. Although I study and teach many intellectual and cultural traditions, I consider it the highest honor to be able to study and teach the cultural and intellectual traditions of black folk.

The vocation of indulging the life of the mind is just as important as the ingenious accomplishments of basketball heroes and superstar singers, talk show hosts and movie stars. The language of the academy is crucial because it allows me to communicate within a community of scholars whose work contributes to the intellectual strength of our culture. Of course, scholars are often criticized for our highly theoretical languages and our jargon-bloated vocabularies. Some of those criticisms are on target, while others fail to account for the existence of various audiences who speak a variety of languages. Not all speech is intended for general consumption.

While scholars must always be conscious of the ethical consequences of what we think and write, the sheer freedom to pursue ideas where they honestly lead us should never be compromised. The language of the academy is most important to me because it provides a critical vocabulary to explore the complex features of American and African-American thought and life. The language of the academy should never divorce itself from the political crises, social problems, cultural circumstances, moral dilemmas or intellectual questions of the world in which we live.

This is why I find my role as a radical democrat and public intellectual a crucial one. As a radical democrat, I am committed to the idea that the political, economic, and cultural resources of our nation should be fairly distributed according to principles of justice and equality. The enormous suffering of many people—but especially working-class and poor peoples—is the most cogent argument for us to adopt the principles of radical democracy in political life. People of color fall disproportionately within the ranks of the suffering. I find the language of radical democracy—with its stress on the integrity of resource sharing, the equitable allocation of goods like education and employment, and the downward redistribution of material wealth—a compelling and useful language with which to analyze and actualize political behavior.

As a public intellectual, I am motivated to translate my religious, academic, and political ideas into a language that is accessible without being simplistic. As a public intellectual, I am committed to the belief that millions of Americans, including black folk, hunger for clear explorations of complex problems. Clarity, unquestionably, is a relative quality that is determined by the nature of the audience being addressed. In the main, however, public intellectuals bring their considerable critical resources to bear on problems of public life that demand complexity of vision, economy of expression, clarity of thought, and rigor of analysis.

As a public intellectual, I think I have a responsibility to encourage my diverse audiences—whether on the op-ed pages of *The New York Times* or on the *Today* show—to consider viewpoints they may deem unpopular, naive, or ludicrous. The public intellectual must be just that: an *intellectual* thinking out loud in various *publics*. Whether the issue is gangsta rap or affirmative action, the presidential budget or O. J. Simpson, AIDS or literary theory, the public intellectual must be willing to think hard about problems without being embarrassed by his vocation of critical consciousness or being afraid to render the results of her learning in lucid language.

As I confront the sometimes conflicting demands of the diverse audiences I speak to and from, considerable tensions arise. Many of my fellow radical democrats are skeptical of my religious commitments. My co-religionists are often wary of my political beliefs. Many of my academic colleagues are suspicious of my public intellectual activities. The rewards of such activities, however, are deep and lasting. I get real joy in writing about R&B and hip-hop for *Vibe Magazine* and *Rolling Stone*, and about the black family for *The Washington Post*. I get real pleasure in preaching at Riverside Church and at Tabernacle Missionary Baptist Church to congregations hungry for affirmations of love in the midst of hate and hope in the face of death. I derive great satisfaction from a lecture delivered to my class at the university, or at colleges I speak to around the nation, especially as students are either motivated or enlightened by my labor. I am uplifted when I speak to union members or factory laborers who feel inspired to work even more diligently for justice and equality.

I am, I suppose, an inveterate polyglot, a skill driven as much by necessity as by pleasure. The various tongues I have engaged—preacher, professor, political activist, and public intellectual—remind me that at our best human beings are constantly going *between* different idioms, vernaculars, vocabularies, even voices, in seeking to understand ourselves and the cultures we create. It is this state of *betweeness* that is at once excruciatingly painful and exhilarating.

As I look at black cultures in general, they are shaped and defined *between* several forces: between the secular and the sacred; between prosperity and poverty; between the artistic and the empirical; between the lofty and the lowly; between the immediate and the deferred; between the tragic and the comic; and between the specific and the universal. From where I stand now—as a committed preacher and public intellectual constantly analyzing and addressing the crises of black life—black culture is constantly being redefined between the force of religious identities and secular passions. Somewhere between God and gangsta rap.

As I write, albums by hardcore rapper Tupac Shakur and black pop legend Stevie Wonder hold the top two positions on the music charts. Shakur raps about the vicious consequences of being a young black male in the ghetto. Wonder sings about being made brand new by the love God has brought him. But either artist is as easily able to speak to subject matters usually associated with the other. Such a state of black betweeness and border crossing is surely worth investigating. And the many tongues through which black culture speaks and is interpreted are certainly worth celebrating. Our lives depend on it.

CONTENTS

1

Invocation

Letter to My Brother, Everett, in Prison

Dear Everett:

How are you? I suppose since we've talked almost nonstop on the telephone over the last five years, I haven't written too often. Perhaps that's because with writing you have to confront yourself, stare down truths you would rather avoid altogether. When you're freestyling in conversation, you can acrobatically dance around all those issues that demand deep reflection. After five years, I guess it's time I got down to that kind of, well, hard work, at least emotionally and spiritually.

I've been thinking about you a lot because I've been talking about black men quite a bit—in my books, in various lectures I give around the country, in sermons I preach, even on Oprah! Or is it the other way around, that I've been talking about black men *because* I've been thinking about you and your hellish confinement behind bars? I don't need to tell you—but maybe I'll repeat it to remind myself—of the miserable plight of black men in America.

I am not suggesting that black women have it any better. They are not living in the lap of luxury while their fathers, husbands, brothers, boyfriends, uncles, grandfathers, nephews, and sons perish. Black women have it equally bad, and in some cases, even worse than black males. That's one of the reasons I hesitate to refer to black males as an "endangered species," as if black women are out of the woods of racial and gender agony and into the clearing, free to create and explore their complex identities. I don't believe that for a moment.

I just think black women have learned, more successfully than black men, to absorb the pain of their predicament and to keep stepping. They've learned to take the kind of mess that black men won't take, or feel they can't take, perhaps never will take, and to turn it into something useful, something productive, something toughly beautiful after all. It must be socialization—it certainly isn't genetics or gender, at least in biological terms. I think brothers need to think about this more, to learn from black women about their politics of survival.

I can already hear some wag or politician using my words to justify their attacks on black men, contending that our plight is our own fault. Or to criticize us for not being as strong as black women. But we both know that to compare the circumstances of black men with black women, particularly those who are working-class and poor, is to compare our seats on a sinking ship. True, some of us are closer to the hub, temporarily protected from the fierce winds of social ruin. And some of us are directly exposed to the vicious waves of economic misery. But in the final analysis, we're all going down together.

Still, it's undeniable that black men as a whole are in deplorable shape. The most tragic symbol of that condition,

I suppose, is the black prisoner. There are so many brothers locked away in the "stone hotel," literally hundreds of thousands of them, that it makes me sick to think of the talent they possess going to waste. I constantly get letters from such men, and their intelligence and determination is remarkable, even heartening.

I realize that millions of Americans harbor an often unjustifiable fear toward prisoners whom they believe to be, to a man, unrepentant, hardened criminals. They certainly exist. But every prisoner is not a criminal, just as every criminal is not in prison. That's not to say that I don't believe that men in prison who have committed violent crimes can't turn around. I believe they can see the harm of their past deeds and embrace a better life, through religious conversion, through redemptive social intervention, or by the sheer will to live right.

The passion to protect ourselves from criminals, and the social policies which that passion gives rise to, often obscure a crucial point: thousands of black men are wrongfully imprisoned. Too many black men are jailed for no other reason than that they fit the profile of a thug, a vision developed in fear and paranoia. Or sometimes, black men get caught in the wrong place at the wrong time. Worse yet, some males are literally arrested at a stage of development where, if they had more time, more resources, more critical sympathy, they could learn to resist the temptations that beckon them to a life of self-destruction. Crime is only the most conspicuous sign of their surrender.

I guess some, or all, of this happened to you. I still remember the phone call that came to me announcing that you had been arrested for murder. The disbelief settled on me heavily. The thought that you might have shot another man to death emotionally choked me. I instantly knew what E. B. White meant when he said that the death of his pig caused him to cry internally. The tears didn't flow down his cheeks. Instead, he cried "deep hemorrhagic intears." So did I.

Even so, a cold instinct to suspend my disbelief arose, an instinct I could hardly suppress. I was willing, *had* to be willing, to entertain the possibility that the news was true. Otherwise I couldn't offer you the kind of support you needed. After all, if you really had killed someone, I didn't want to rush in to express sorrow at your being wrongly accused of a crime you didn't commit. Such a gesture would not only be morally noxious; it would desecrate the memory of the man who had lost his life.

If I wasn't able to face the reality that you might be a murderer, then I would have to surrender important Christian beliefs I preach and try to practice. I believe that all human beings are capable of good and evil. And regarding the latter, wishing it wasn't the case won't make it so. Too often we deny that our loved ones have the capacity or even inclination for wrongdoing, blinding us to the harm they may inflict on themselves and others.

I eventually became convinced that you were innocent. Not simply because you told me so. As one lawyer succinctly summarized it: "To hear prisoners tell it, there are no are guilty prisoners." After discerning the controlled anger in your voice (an anger that often haunts the wrongly accused) and after learning that the police had discovered no weapon, motive, or even circumstantial evidence, I believed you were telling me the truth. Plus, you had been candid with me about your past wrongdoing. And in the wake of your confessions of guilt, you repeatedly bore the sting of my heated reproach. For these reasons, I believed you were not guilty.

I realized then, as I do now, that these are a brother's reasons. They are the fruit of an intimacy to which the public has no access and in which they place little trust. Many of the reasons that led me to proclaim your innocence are not reasons that convince judges or juries. Still, I felt the bare, brutal facts of the case worked in your favor. A young black man with whom you were formerly acquainted was tied up in a chair on the second floor of a sparsely furnished house. He had tape tightly wrapped around his eyes. He was beaten on the head. He was shot twice in the chest at extremely close range, producing "contact wounds."

After breaking free of his constraints, he stumbled down the flight of stairs inside the house where he was shot. Once he made it down the stairs outside the house, he collapsed on the front lawn of the house next door. As he gasped for breath while bleeding profusely, he was asked, first by neighbors, then by relatives who had arrived on the scene, and later by a policeman, "Who did this to you?" Something sounding close enough to your name was uttered. The badly wounded man was pronounced dead a short time later after being rushed to an area hospital.

In the absence of any evidence of your participation, except the dying man's words, I thought you'd be set free. After all, he could be mistaken. Given the tragic conditions in which he lay dying, he might not have had full control of his faculties. Was his perception affected by his gunshots? Was his mind confused because of the large amounts of blood he had lost? Unfortunately, there was no way to be certain that he was right. There was no way to ask him if he was sure that you were one of the culprits (he said "they" a couple of times) who had so barbarically assaulted him. But without his ability to answer such questions, I believed there was no way you would be imprisoned. Surely, I thought, it took more than this to convict you, or anyone, of murder.

I was wrong. The murdered man's words, technically termed a "dying declaration," were admitted into court testimony and proved, at least for the jury, to be evidence enough. I was stunned. In retrospect, I shouldn't have been. Detroiters were fed up with crime, including the ones who peopled the black jury that convicted you. How many times had this apparent scenario been repeated for them: black men killing other black men, then seeking pardon from blacks sitting on a jury in a mostly black city?

When it came time to sentence you, the judge allowed me to say a few words. I felt more than a little awkward. Although I didn't believe you were guilty, I knew that if I said so the judge would ignore my presentation. In his mind, the jury had settled the issue of your culpability. I didn't know how much I should refer to your past, or to the social forces that shape human action. I figured that the last thing I needed to do was sound like a hot-shot intellectual trying to enlighten the masses.

I knew in my heart that I shouldn't avoid mentioning those beliefs I held to be true, for instance, that economic misery can lead to criminal activity. At the same time, I didn't want to be mistaken for defending the belief that social structures alone determine human behavior. I also wanted to avoid inflicting any more pain on the murdered man's family, most of whom believed that you were guilty as sin. And I didn't want to be condescending. I didn't want to sound like the brother who was righteous, who had made it good, making excuses for the brother who had gone completely wrong. I wanted to speak from the heart, so I didn't use a prepared text. I wonder if you remember what I said?

> Your Honor, I'm a minister of the gospel and I'm also a scholar, and a teacher at a theological seminary. I, of course, want to express first of all my deep sympathy to the family of the man who died. They have endured enormous hurt and pain over this past year. I want to say to you in my brief remarks that I am deeply aware, in an ironic sense, of why we're here. Sentencing is a very difficult decision. I have been deeply committed over the years to justice in American culture and also to examining the workings of the legal system.

> On the other hand, I also understand the societal forces—such as poverty and joblessness and structural unemployment and limited social options and opportunities for legitimate employment—that many people of our culture, particularly black men, face. It is also ironic that I'm here because I write in my professional life about . . . social forces which often leave young black men feeling they have no other options but to engage in . . . criminal activity in order to sustain their lives. Unfortunately many make that choice.

I grew up in the urban poverty of Detroit, as did the other members of my family. Therefore I understand not only from a scholarly viewpoint, but from a personal viewpoint, limited life options and the kind of hopelessness and social despair they can breed in a person.

I come here this morning pleading and praying for leniency in my brother's case. As his lawyer has already stated, the mystery that surrounds the events of that day continues to prevail. In any regard, I can attest to my brother's character, that he is not a hardened criminal. He has made unwise choices about the activity of his life in the past. He has made choices which have encouraged him to engage in a lifestyle that I'm sure at this point he is not proud of. At the same time I think this penalty far exceeds any crime that he has been involved in.

Above all, my brother is, I think, ripe for a productive future in our society. Although he has indeed made a noticeable change in jail, a prolonged stay in prison, I feel, will not greatly contribute to any sense of rehabilitation that the Court might think prison offers.

Unfortunately the prisons of our land often reproduce the pathology that they seek to eliminate. Because of his own poor beginnings in our city, the death of our father at a crucial time in his life, and because he's been subjected to the forces I've already referred to, my brother has made poor choices. But he's also shown a remarkable strength of faith and renewed spiritual insight. He's shown a remarkable sense of concern . . . about all the people involved in this case and not just himself. . . . In conclusion, Your Honor, I would plead and pray that . . . your deepest discretion and most conscientious leniency prevail in your sentencing of my brother this morning.

I have rarely been more depressed, or more convinced that my words meant absolutely nothing, than when the judge's words, all-powerful words, revealed your future. Life. In prison. An oxymoron if I've ever heard one.

You have managed to squeeze an ounce of invention, or

should I say, self-reinvention, from the pound of cure that prison is said to represent. When I first learned of your new identity, Everett Dyson-*Bey*, I was neither dismayed nor surprised. Frankly, my position is simple: do whatever is necessary to maintain your safety and sanity in prison without bringing undue harm to another person. You're a strong, muscular fellow, and I didn't think you'd have much trouble staying safe. I took your change of religions—from the Christianity you inherited as a child to the Moorish Temple Muslim belief of your new adulthood—to be an encouraging, even creative defense of your sanity.

I am disappointed, though, by the response of the black church to your predicament. I suppose since I've been to see you countless times over the last five years, it could be claimed that those visits count for my church's mission to those locked away. But we both know that's bogus. That line of reasoning insults the integrity and slights the example of so many who've followed Jesus in "visit[ing] those who are in prison." You haven't been visited a single time in prison by anyone visiting *as* a Christian minister, or *as* a concerned church member. Thank God our pastor visited you in jail before you went to prison. But the church is larger than him. I spoke several times to the minister in charge of prison visitation about going to see you. My requests were futile.

I don't know why so many black Christians avoid the prisons. Of course, I realize that hundreds of black churches have prison ministries that make a real difference in inmates' lives. But the average pew sitting member, or for that matter, the regular church minister, rarely gets into the thick of prison life in the same way, say, as members and ministers of the Nation of Islam. Or the Moorish Temple. Perhaps it has something to do with how black Muslims with smaller numbers than black churches must proselytize when and where they can. Since many of their members have served time, they may be more willing to reach back to help those left behind. Then, too, the application to prisoners' lives of the stringent ethical code taught in black Muslim settings often brings welcome relief to the moral chaos into which so many inmates have descended.

Another reason for their success may be that black Muslims take seriously their theological commitments to racial uplift and reconstruction, especially among the poor and imprisoned who are most in need of that message. Perhaps it's a simple class issue. The more legitimacy some black Christian denominations gain, the higher class status they acquire, the less they appear inclined to take care of "the least of these." In the end, I'm glad you've discovered in the Moorish Temple what you couldn't find—or perhaps what couldn't be found in you   through Christian belief.

Many people think the sort of religious change you have experienced is a "foxhole" conversion, a transformation brought on by desperate circumstances that will be rejected as soon as you're set free. That may be the case. If it is true, you certainly won't be the first person it has happened to. But hold on to the hope your religion supplies as long as you can. There will be other desperate situations after you leave prison. Besides, so-called normal religious people experience a series of crises and conversions over the years in settling down to a deeper faith. Even those folk who don't walk through its doors every time church opens often have meaningful conversations with God.

I think our father was one of those people. He was a complex man who worked extraordinarily hard and who believed deeply in God. But he wasn't very religious, at least not in any traditional way. When people discover I'm a Baptist preacher, they often ask if preachers run in my family, if my father was a preacher. I laugh inwardly, sometimes out loud, thinking of what an odd image that is, Daddy as a preacher. It's not that he cussed like a sailor. I know too many preachers who do that as well. And its not because he had a short fuse. So do most prophets, biblical and current ones too. I guess it's their righteous rage at evil, their ill-tempered tirades translated as holy damnation. But the line between their baptized fussiness and plain old invective is sometimes quite thin.

I think what causes my bemused response is Daddy's genuine humility. Most preachers I know aren't that humble. I don't think that's all bad. Many can't afford to be. The tribulations of their office are enough to shatter a fragile ego. But the annoying hubris found in so many ministers was completely absent in Daddy. Yet this humble man also displayed ferocious anger which frightened me. True, it didn't last long when it surfaced. But its concentrated expression had devastating consequences. And often—I think too often—it had its most harmful effects on his children's behinds—not to mention on their minds.

To be honest, I don't completely understand why Daddy so readily turned to the strap to discipline us. Perhaps he was treated the same way when he was a child. Maybe the humiliations he suffered didn't have any other outlet. I remember once when he and I were working for "Sam's Drugs" as janitors. I was in my middle teens, which meant Daddy was in his late fifties. A light fixture had been broken in the ceiling of the drugstore. In order to reach it, Daddy climbed a step ladder that I was holding as Ben, the Jewish owner, looked on. When Daddy misstepped and slipped down a couple of rungs on the ladder, Ben became angry.

"Oh, Everett," he indignantly declaimed. "You're just like a little boy. Can't you do anything right?"

Daddy didn't say a word. I was so mad at Ben, and humiliated for Daddy at the same time. I remember thinking of how strong Daddy was, how physically domineering he could be. Yet none of that mattered as Ben reduced his humanity, and as I interpreted it then, attacked Daddy's manhood as well. If I felt that as a teen, what did Daddy feel? Where did he put that anger? Is that at least part of the reason he let his rage loose on us?

As you know, the debate about corporal punishment is

raging in our nation. There used to be a belief that there was a racial divide on these matters, at least when we were growing up. Black folk in favor, white folk opposed. Even though I don't think it's that simple (where one lives, either in the city or the suburbs, and one's class identification, are important too). I don't deny that racial differences exist.

Recently, though, I think the gulf between black and white views on child rearing has probably narrowed. A new generation of black parents has questioned and often rejected the wisdom of whipping ass. To be sure, you still hear black folk saying, "The problem with white folk is that they let their kids get away with murder, let them talk and act any way they want to without keeping them in check." You also hear black parents and the experts they listen to arguing that corporal punishment encourages aggressive behavior, stymies the development of moral reasoning, hinders self-esteem, and even causes children to be depressed. No such theories prevailed in our household.

I must admit, I tend toward the newfangled school of thought, even though I haven't always put it into practice now that it's my turn to parent. In fact, during your nephew Michael's childhood and early adolescence, I didn't know anything about "time out." As a teen father, I had barely survived the pain of my own rearing and the violence I'd encountered. I knew what I saw, repeated what was done to me. And I regret it.

One of the most painful moments I experienced involving punishment occurred when I was a teacher and assistant director of a poverty project at Hartford Seminary. Brenda (then my wife), my son Mike, and I were in our car as I drove to work to pick up some papers one evening. Down the street from the seminary, Mike had behaved so badly in the car that I pulled over to the side of the road to discipline him—three licks on his hands. In my view, it was a very light and well-deserved spanking. After administering this punishment, I drove the single block to the seminary.

Before I could park my car in front of the seminary two white policemen drove up in a squad car. They got out of the car and one of the policemen approached my door, instructing me to get out of the car. His partner walked up to Brenda's side of our car.

"Can I ask you why you're stopping me, officer?" I asked politely and professionally. I'd learned to do this, as most black men in America have learned, to keep the blue wrath from falling on my head.

"Just get out of the car," he insisted.

As I got out of my car, I informed the policeman that I worked at Hartford Seminary.

"I'm a professor here," I said, pointing to the seminary behind me.

"Sure," the policeman shot back. "And I'm John Wayne."

The policeman instructed me to place my hands against the car and to lean forward. I knew the drill. I'd done it too many times before. I could hear the other policeman asking Brenda if everything was alright, if my son was harmed. Mike was in the back seat crying, afraid of what the police were going to do to me.

"I'm fine, I'm fine," Mike cried. "Why are you doing this to my Dad?"

From the pieces of conversation I heard between the second cop and Brenda, I gathered that someone—a well-meaning white person no doubt—had spotted me spanking Mike and reported me as a child abuser.

Just as Brenda told the cop how ridiculous that was, two more police cars rolled up with four more white men. "Damn," I thought, "if I had been mugged, I bet I couldn't get a cop to respond within half an hour. And now, within five minutes of spanking my son, I've got six policemen breathing down my neck."

As the other cops surrounded our car, the policeman hovering over me refused to explain why he stopped me. He forcefully patted me down as we both listened to Brenda and Mike explain that nothing was wrong, that Mike was fine.

"You sure everything's alright?" the cop talking to Brenda asked once again for degrading emphasis. She angrily replied in the affirmative.

Finally my knight in shining armor spoke to me.

"We got a complaint that someone was hurting a child," he said.

"I can assure you that I love my son, and that I wasn't hurting him," I responded in a controlled tone.

"I spanked my child now so that he wouldn't one day end up being arrested by you."

"We have to check on these things," the second cop offered. "Just don't be doing nothing wrong."

He shoved me against the car to make his point. With that, the six cops got back into their cars, without apology, and drove off.

I don't have to tell you that the situation was utterly humiliating. I resented how I'd been treated. I felt the cops had deliberately intimidated me. They embarrassed me in front of my family under the guise of protecting them. I think their behavior is fairly typical of how many white men with authority treat black men. They are unable to be humane in the exercise of power. They run roughshod over black men in the name of serving a higher good, such as protecting black women and children from our aggression. The irony of course is that white men ignore how their violence against black men has already hurt millions of black families, including black women and children. In fact, the effect of much of white male hostility is not to help black women and children but to harm black males. Fortunately for me, Brenda and Mike understood that truth. Neither of them trusted the cops' motives for a moment.

Still, the incident forced me to imagine the impact my punishments had on Mike. I thought about how he might interpret the discipline I gave him. I wondered how spankings made him *feel*, despite the reassurances of love I prefaced to any punishments I gave him. The irony, too, is that I was reading social and cultural theorists who were writing about discipline and punishment. While I found many of them extremely enlightening about big social forces and how they molded people's habits of life at home and in the world, I sometimes wondered if they had any children. I continued to talk to Mike about these matters, apologizing to him about my past disciplinary practices, promising him, and mostly living up to it, that I would look for alternatives to physical punishment.

Of course Daddy lived in a world where such considerations were impossible. If you don't control your kids, they'll control you. That's the logic that informed his decisions. If you don't beat their asses, they'll beat yours one day. I guess depending on where you stand on such issues, the rash of recent slayings of parents by their children either proves or undermines such a theory. In any case, I eventually grew to hate Daddy for the violence of his punishments. I can still hear him saying "get me that 'hind pepper'," referring to the quarter-inch-thick, twelve-inch-long piece of leather he used to whip us. Occasionally, he'd plant his size twelve foot right up my posterior.

I know, of course, that no one on our block would have called that child abuse. And neither did I. Given the black cultural logic of the time during which he was reared, and during which he and Mama reared us, Daddy was simply attempting to keep his brood in line. (What we must not forget is that during an earlier time in our nation, black folk beat their children at home so they wouldn't give white men lip in public. If a black child wasn't strictly disciplined, he might say or do something that might cause him untold danger away from the protection of parents. Even though that logic may be long exhausted, some habits die hard.)

My resentment of his whippings got so bad that he once told Mama that he thought I hated him because he wasn't my biological father. When Mama told me that, I was crushed.

For despite his discipline, I knew he loved me as if I was his very own, like I was your full blood brother. For that reason, I have never made the distinction between any of us five boys who came up together. In my mind, not only did we have the same mother, but we shared the same father. He was as much father as most of my friends had, and often, much, much more. Since he adopted me when I was two, he is the only father I have ever known. He was Daddy to me, just like he was to you.

No, I was very specific about my beef with him. It wasn't blood, it was those beatings. The same ones he gave to you, Anthony, Gregory, and Brian. And probably to John Everett, Etta James, Robert, and Annie Ruth, our late brothers and sisters from Daddy's previous marriages.

My conflict with Daddy came to a head when I was sixteen, the same age Mike is now. He had ordered me to do something, what I can't remember. I do remember feeling the familiar threat of physical punishment behind his words if I didn't immediately obey. I had had enough. We were at the house, upstairs on the second floor. He barked his orders, but I wasn't moving fast enough.

"Move, goddammit, when I speak to you," he bellowed.

The resentment weighed me down, and slowed my legs. I knew instantly that we were heading for a showdown. Daddy jumped up from the bed in his room and moved toward me. Even that gesture failed to speed my pace. This wasn't worker slowdown, a domestic uprising against an unjust guardian. This was sheer frustration, anger, and weariness.

"Move, I said," Daddy repeated. I didn't.

Then he grabbed me by the arm and pushed me against the wall. Something in me exploded. Or did it snap? Either metaphor, or perhaps both of them, captured my state of mind, my state of soul.

"Fuck it, man," I heaved. "You just gonna have to kill me, 'cause I refuse to be scared any more."

I guess he took me seriously. He literally lifted me off the ground with his left arm, his massive chocolate hand sunk deep into my yellow neck as he pinned me against a hallway wall. They didn't call him "Muscles" for nothing. I thought for sure that he might really kill me. I didn't care anymore. I was tired of running. Mama saved me.

"Everett," she hollered. It was all she said. But it was enough to bring Daddy to his senses, to make him drop me to the ground before he completely choked me. Never mind my gasping. I felt free, delivered of some awful demon of fear that no longer had power over me. It was my emancipation proclamation and declaration of independence all rolled into one moment. It was a milestone in my relationship to Daddy.

For the next seven years, his last on earth, Daddy and I got along much better. After I got Terry pregnant at eighteen and married her, and after Mike was born, Daddy and I grew much closer. In fact, he'd often cook for me and Terry because we were so poor at times that we didn't eat every day. In fact, at times, we didn't eat for two or three days in a row. But then we'd go by the house, and Daddy would always give us a good meal. I even sent Daddy a Father's Day card in 1981 when I was in Knoxville attending college. I told him how much I loved him, and how much I appreciated the fact that we had overcome our differences now that I was a man with major responsibilities. A few weeks later, he was dead from a heart attack at sixty-six. So young when you really think about it.

But I must confess, even now as a thirty-five-year-old man I have dreams of Daddy doing violent deeds to me, whipping me in vicious ways. The lingering effects of the whippings Daddy administered are illustrated in a story I heard about a boy and his father, who sought to rid his son of his habit of lying. The boy's father hammered nails into a piece of wood for each lie his son told. Finally, when the board was nearly full, the boy pledged to stop lying. And his father promised to pull a nail out each time his son told the truth. When the board was completely empty, the boy began to cry.

"What's wrong, son?" the boy's father asked. "You should be happy. You've stopped lying, and the nails are all gone."

"Yes," the boy replied. "The nails *are* gone, but the holes are still there."

Well, the holes are still there for me as well. My psyche bears the marks of spiritual and psychological violence. But I am not bitter towards Daddy. I honestly believe he was a good man trying to do his best in a world that was often difficult for him. The older I get, the more clearly I understand the forces he faced.

I guess I'm sharing all of this with you now because we never enjoyed this kind of intimacy before your imprisonment. A shame, but it's true. And even though we grew up in a household where we knew we were loved, we rarely, if ever, heard the words, "I love you." Daddy taught us to be macho men, strong enough to take care of ourselves on the mean streets of Detroit. And though Mama protested, thinking Daddy was trying to make us too rough at times, I'm sure we both appreciate many of his efforts to prepare us for an often cold-hearted, violent world.

I yearned for a home where we could be both strong and vulnerable, tough but loving. Daddy's reading of the world led him to believe it was either one or the other. He chose to teach us how to survive in a city that was known then, in the seventies, as the "Murder Capital of the World." And because I loved books, and not the cars that you and Daddy and Brian loved to work on, he sometimes thought I was "too soft." Daddy was really proud of me later when I excelled at school. He wanted me to be better than he was.

I remember once when I was about eight years old, I was mimicking his pronunciation of the number 4. He pronounced it "foe." I followed suit. But he stopped me.

"Don't you go to school, boy?" he asked.

"Yes," I replied.

"Don't you know how to say that right?"

"Yes."

"Then do that from now on. Okay?"

I've never forgotten that exchange. He didn't have a great

education, but he sure wanted me to be learned. Indeed, he wanted the best for all his boys. I imagine if he was alive he'd be heartbroken that you're in prison. Daddy was the complete opposite of so much of what prison stands for. He rose every day before dawn, even after he retired from the factory, and worked until evening, cutting grass, laying sod, painting, or working as a maintenance man. I learned my work ethic from him. I can still hear him saying, "Boy, if you gonna do a job, do it right or don't do it at all." I've repeated that to Mike at least a million times. And of course, his other famous saying was "If you start a job, finish it." That is, other than his maxim: "Laziness will kill yo' ass."

And even when he worked those thirty-three years at Kelsey-Hayes Wheelbrake and Drum Factory, he often put in sixty or seventy hours. I swear I once saw a stub where he had worked nearly eighty hours, pulling a double shift for an entire week. It was Daddy's example that led me to work two full-time jobs after Terry got pregnant with Mike. (He warned me then, "The more money you make, the more you spend." He was right, of course.) I'd go to a maintenance job from 1 A.M. to 7:30 the next morning, and then work a menial "construction job" (a misnomer, to be sure) from 7:30 to 4:30 in the evening. And I still had to get food stamps while Terry was enrolled in WIC (Women, Infants and Children!). That stuff saved our lives.

I'm glad that you and I have learned to talk. To communicate. To express our love for one another. It hasn't been easy seeing you cooped up like an animal when I visit you. But the one good thing to come out of all of this is that at least we're getting to know each other better. That's why I feel good about telling the world about you.

Even as I talk about you on television and radio, though, I always try to impress on the audiences and interviewers in the short time I have that ours is no "one son makes good and the other makes bad: what a tragedy," scenario. I'm not trying to pimp your pain or commercialize your misery to make a name for myself. That's because I believe in my heart, and I hope you do too, that it could just as easily be me in your cell. I don't want people using our story as a justification for rewarding black men like me who are able to do well while punishing brothers like you who've fallen on harder times.

No matter how much education I've got, this Ph.D. is no guarantee that I won't be treated cruelly and unjustly, that I won't be seen as a threat because I refuse to point the finger at "dem ghetto niggers" (a statement made by black and white alike) who aren't like me. I'm not trying to erase class differences, to pretend there's no difference in a black man with a Ph.D. and a black man who's a prisoner. I'm simply saying I can't be seduced into believing that because I've got this degree I'm better.

How could I be? I was one of "dem ghetto niggers" myself. Even now I think of myself as a ghetto boy, though I don't live there anymore, and I refuse to romanticize its role in its inhabitants' lives. Not even survivor's guilt can make me that blind. But being from the ghetto certainly leaves its marks on one's identity. Don't get me wrong. I'm all for serious, redemptive criticism of black life at every level, including the inner city. There's a difference between criticism that really helps and castigation that only hurts.

I should close this letter for now. I fear I've touched on many sensitive spots, and you may sharply disagree with some of the things I've written. But that's alright. The important thing is that as black men, as black brothers, we learn to embrace each other despite the differences that divide us. I hope you write me back. I'd really like to know what you think about what I've said. In the meantime, stay strong, and stay determined to renew your spirit and mind at the altar of devotion to God and our people. In the final analysis, it's the only thing that can save us all.

Peace and Love,
Mike

TESTIMONIALS

the joys and concerns of black men's lives

2

Obsessed with O. J.

Meditations on an
American Tragedy

When O. J. Simpson took that long, slow ride
down the L.A. freeway in A. C. Cowlings' bronco,
it wasn't the first time he used a white vehicle to
escape a black reality. I'm not referring to interracial marriage per se. Compatibility in love doesn't
respect race, height, sex, color, age, culture, religion, or nationality. Transgression and affection
often team up to knock down artificial conventions built on bias and ignorance. But sex between
blacks and whites is an especially volatile instance
of interracial intimacy. Every gesture of crossover
and exchange between blacks and whites indexes
the bitter history of American race. When many
interracial couples forge ahead against social

taboo, they often act courageously to undermine rigid racial beliefs.

Some blacks, though, pursue white lovers and lifestyles in a way that only reinforces the rules of race. White identity signifies for them the desirable, the healthy, the stable. Black identity, by contrast, symbolizes the undesirable, the unhealthy, the unstable. Although I cannot know for sure, I imagine that O. J., however unconsciously, may subscribe to these beliefs. He apparently belongs to that fraternity of black men who "have-to-have-a-white-woman-at-all-costs." There's a difference, after all, between preference and obsession. (Of course, even preferences don't jump at us out of a cultural void; our deepest desires bear the imprint of the society that shapes them.)

And in O. J.'s case, it seems that learning to speak correctly meant learning to talk like white folk. (I'm not equating excellence in speaking or writing "standard" English with "acting white.") He wanted to get as far away from his ghetto roots as his legs, wealth, fame, and diction could carry him. Good for him. But because he appears so uncomfortable with the idea that you can be identifiably black and have all those things, he has drowned his racial identity in an ocean of whiteness.

Critics have argued bitterly about what light O. J.'s case will shed on race in America. That's not entirely clear. Unquestionably, the case has all the elements of national farce and intrigue. I'm not sure if it will affect the lives of everday black folk. As he faces the possibility of being locked away, O. J.'s abrupt decline embodies the plight of more black men now than when he achieved heroic heights. (After all, there are hundreds of thousands more black men in prison than in the Hall of Fame.) Less than a week before his fall, O. J. appeared to be a lifetime away from the heartless siege of troubles that vex millions of black men.

His current condition has added O. J.'s name to a growing list of (in)famous black men whose personal problems have made them poster boys for the perversions of patriarchal culture. Mike Tyson and date rape. Clarence Thomas and sexual harassment. Michael Jackson and child molestation. And now, O. J. Simpson and spousal abuse. Each of these problems merits serious action, and these men, if guilty, should be held responsible and punished accordingly. But these maladies are ancient. How is it that these black men have managed to do in disrepute what most black men can't do in honest achievement: transcend race to represent America?

Make no mistake. O. J. is not Rodney King. In the racial firmament, the King case can be considered a supernova, illuminating the ground of race relations beneath its harsh but powerful light. Millions of poor black men can identify with Rodney King because police brutality is a staple of their adolescence and adulthood, a ritual of initiation into a fraternity of black male pain. Despite his working-class hustling roots, the meaning of Rodney's beating could nevertheless travel in an upwardly mobile fashion: even well-to-do black males understood King's horror because it could be directed toward any black man on any given day.

O. J.'s case, by contrast, is considerably more narrow. Most black men charged with a capital crime have little money to seek responsible representation. Most do not count the police as their friends, or even lackeys. Nor do they have the use of fame as a powerful deterrent to their conviction. If Rodney King is an exploding star, it may be that O. J. Simpson is a black hole, a collapsed star of such immense gravity that no light can escape.

Johnnie Cochran has been called a modern-day Joe Louis. In part, I can see that. He has fought tough legal battles for some of our most beleaguered black brothers: Jim Brown, Todd Bridges, Michael Jackson, and now O. J. Rascals all, in their own ways. In representing them, Cochran has slugged it out with a justice system that often punishes black men with frightening consistency.

Unlike the Brown Bomber, though, Cochran's gifts spill forth from his golden throat. He is smooth and silky, an orator of great skill whose rhetoric reflects his Baptist roots and his early days as an insurance salesman. He performs the law, dramatizing its arcane rituals of argument and translating its esoteric dogmas into stirring, poetic declaration. For many blacks Cochran *is* the law, masterfully taming the chaos of white contempt camouflaged in legal language and protected by obscure codes and regulations.

The pride so many blacks feel in Cochran's performance has a lot to do with an ancient injury to black self-esteem that not even Joe Louis could relieve: the white challenge to black intelligence and its skillful defense in eloquent black speech. Among his many racial functions, the black orator lends credence to claims of black rationality. When black folk in barbershops and beauty salons say of Cochran that "The brother can talk," what they mean in part is that the brother can think. Thinking and speaking are linked in many black communities. And neither are abstract reasoning and passionate discourse often diametrically opposed in such circles. Like all great black rhetoricians, Cochran makes style a vehicle for substance.

When *Time* magazine blackened O. J.'s face on its cover, it was a gesture full of irony and, yes, dark humor. *Time*'s act raises several questions. Was their artificial enhancement of O. J.'s natural hues, which forced us to be more conscious of his color, a signifying move, suggesting that O. J. needed help with his color consciousness? Was it a subversive move, motivated by *Time*'s hidden ties to skin nationalists who argue a link between pigment and personality? Were they demonizing him by darkening him, making O. J. a Darth Vader where many once believed he was a Luke(warm) Skywalker? Was *Time* trying to help a brother out by boosting his melanin count to swing public sentiment his way? Or were they vilifying O. J.'s vanilla vision of black identity? Were they extending the spookification of black public faces? Or were they simply doing with O. J. what the media have done to countless other black men: giving him the benefit of the lout?

The riveting and repulsive drama of O. J. Simpson's freakish unraveling before our very eyes contains many ironies. An athlete whose brilliant moves on the football field were marked by beauty and grace now left an international audience aghast at his ungainly flight from the law. A champion who played Prometheus to a nation of Walter Mittys now shrank in stature to a shriveled, self-defeating parody of his former strength. An icon with an ingenious talent for turning gridiron glory into Hollywood fame and fortune was now bedeviled by the media that helped make him a national figure. And a man whose face and initials were broadly familiar became in an instant a stranger with a secret history of spousal abuse that may prove to have been an unseemly rehearsal for murder.

One of the most remarkable features of the initial commentary around Simpson's sad situation is the way in which race, in its deliberate denial, was made even more present. Like Poe's purloined letter, race laid hidden in plain sight.

On the face of things, the denial of race in the Simpson case signaled a praiseworthy attempt by the media to balance its racially skewed reporting of news events. That's not easy when politicians and pundits are obsessed with negatively linking race to everything from welfare reform to crime. But in denying the role of race in the Simpson ordeal, media critics showed that you can't get beyond race by simply pretending it's not there.

The goal should not be to transcend race, but to transcend the biased meanings associated with race. Ironically, the very attempt to transcend race by denying its presence reinforces

its power to influence perceptions because it gains strength
in secrecy. Like a poisonous mushroom, the tangled assump-
tions of race grow best in darkness. For race to have a less
detrimental effect, it must be brought into the light and
openly engaged as a feature of the events and discussions it
influences.

In the case of O. J. Simpson, the fingerprints of race are
everywhere. O. J.'s spectacular rise to fame was aided not only
by his extraordinary gifts, but because he fit the mold of a talented but tamed black man, what was known in his youth as a
"respectable Negro." O. J. received brownie points throughout his playing career as much for who he wasn't as for how
he performed. He wasn't considered, like football star-turned-
actor Jim Brown, a black "buck," an "uppity nigger," an arrogant, in-your-face threat because of his volatile presence and
unpredictable behavior.

From the beginning of his career, O. J. Simpson was mar-
keted to white society as a raceless figure whose charisma
drew from his sophisticated, articulate, public persona. In
this light horrified, disbelieving gasps of "not him" unleashed
at O. J.'s initial public disintegration take on new weight. Let's
face it. The unspoken, perhaps unconscious belief of many
whites is that if he's guilty, if this could happen to O. J.—the
spotless embodiment of domesticated black masculinity—it
could happen to any black man. Translation: no black male
can really be trusted?

At first, the fact that Nicole Brown Simpson was a beautiful
blonde white woman went virtually unremarked upon. Now,
of course, her ubiquitous picture has made it hard not to
notice. The fact that O. J. married Nicole made a lot of people mad.

When a black man marries a white woman, it irks white
supremacists ("he's spoiled one of *our* women"). It grieves

many black mothers ("when a black son brings home a white woman, it's an insult to his mama"). It angers many white men ("she's throwing her life away"). It disappoints many black women ("with all these single black women, why would he choose a white woman?"). It unnerves some white women ("I could never see myself with a black man"). And it raises some black men's ire ("why all these brothers, when they get successful, got to marry a white woman?"). This small sample of anecdotal responses to interracial relationships provides a glimpse of the furious passions and unresolved conflicts that continue to haunt love in black and white.

Were O. J. and Nicole completely immune to such concerns? Probably not. Does the fact that O. J. is charged with killing his *white* wife make a difference in our world? Probably so. Can we seriously doubt that if O. J. had been accused of murdering his *black* wife, and not a symbol of ideal white beauty, we wouldn't be learning of it with a similar degree of intensity, its details so gaudily omnipresent?

Many argue that Simpson's troubles have nothing to do with race, that his fall is instead an *American* tragedy. Of course it is, because all black citizens are Americans, and all of our problems, therefore, are American problems. But we don't have to embrace our American identity at the expense of our race. The two are not mutually exclusive. We simply have to overcome the limitations imposed upon race, to make sure that neither privilege nor punishment are viciously, arbitrarily assigned to racial difference. To erase race is to erase ourselves, and to obscure how race continues to shape American perceptions and lives.

O. J.'s handsomeness has played a large part in his appeal over the years. He has long been the object of the "safe" eroticization of black masculinity by white women. His pretty face is now being beamed everywhere. His facial expressions are deconstructed around the globe. Once the master of his image in a medium where he adroitly projected a cool persona, O. J. has now lost himself, at least his public self, in the infinite gaze of international television. O. J.'s every glance is now filtered through the theories of thinkers with names like Derrida, Sartre, and Foucault. What could he mean by looking up? How could his failure to regard photos of his dead ex-wife possibily signify his guilt or angst? What are the possible meanings of his intense glare at jurors?

Before his demise, O. J. spoke with authority, breaking stereotypes of black sports icons' severe inarticulateness. His precise diction rebutted the vicious subculture of parody that dogged the verbal skills of his boyhood idol Willie Mays. Now O. J. sits mute. He is forced to minstrel his meanings by bucking his eyes like Rochester or Stepenfetchit.

The trial of Colin Ferguson, the convicted Long Island Railroad murderer, briefly gave the Simpson trial a run for its money. Ferguson's case literally captured the lunacy of race. His brilliant and bizarre self-defense was an amalgam of common sense and uncommon senselessness. It was, perhaps, the first postmodern trial, as Ferguson's own fragmented identities competed to both relieve and reinforce the burden of representation. I was haunted throughout his trial by the thought that this was one of the few black persons able to conceive himself in public on his own terms. His performance drew heavily from television, a fascinating pastiche of Perry Mason dramatic flourishes, Matlock rhetoric, and a guilessly imitative collage of Simpson defense team strategies, gleaned, no doubt, from CNN and Court TV.

Ferguson's vicious, destructive bigotry had no greater victim than himself. He is the tragic embodiment of Martin Luther King Jr.'s warnings to blacks of the consequences of unchecked hatred of whites. Ferguson's lethal psychic decline, balanced by runs of extraordinary eloquence, captures as well the truth Grier and Cobb uncover in their masterful book *Black Rage*: when black folk go crazy, it's the delusional mental diseases that bewitch us most. I suppose that's because we so rarely get to be our own complex selves, and when opportunities for self-expression arise, our black identities are often degraded. So the black delusional personality sanctifies excess and distortion, compensating in its sick way for the legitimate strength it is denied by oppression. By its sheer volatility, this form of black insanity calls attention to one of its major causes: white racism.

I was struck, too, by the parallels between Ferguson's demented bigotry and the reign of white lunacy that passed for legal justice in this country during most of this century. How many black lives were sacrificed, how many claims to equality were denied as white judges and jurors cavalierly, callously dismissed overwhelming evidence against white perpetrators? The sheer insanity of this situation was not lost on its black victims. Colin Ferguson's actions are the bitter elaboration of the perverted, insane logic of racial hatred that for too long underwrote our legal system.

For many blacks, including women, the problem of domestic violence in the Simpson case has been subordinated to race. In fact, for many it almost doesn't exist. This denial is achieved by insisting that Nicole was "trash," that she was sexually loose, that she was a party girl who played O. J. for material gain. On this reading, Nicole got what she deserved.

But even if this portrait of Nicole is accurate (and a strong case can be made that it was), it still doesn't justify O. J.'s brutal beating of Nicole. Nor should it have cost her life. After all, a stronger case can be made that O. J. was extraordinarily promiscuous, a man of enormous appetites for varied substances, and beneath the public sheen of class, a brazen bully of women. O. J. and Nicole equally embodied the ugly lives of pretty people.

The truth is that black male violence against black women is a mainstay of relations between the two. The oppressive silence black women have observed in deference to race loyalty, or had imposed on them out of fear, remains a tragically underexplored issue in black life. Domestic violence against women is a concealed epidemic in black communities. It needs to be exposed.

If any good can come out of O. J.'s case for black women, it is that violent behavior should not be tolerated for any reason, racial or domestic. The male sexual ownership of women, the presumption of male discretion over women's bodies that feeds obsession and domination, must simply desist. Plus, sisters, when O. J. walks, he's not coming for you. Does he, or any black man who doesn't display the utmost respect and admiration for black women, really merit such profound loyalty?

The polls continually show that blacks and whites are divided on how they view this case. The wonder is that so many white folk are surprised. Why? Race remains the primary prism through which Americans view reality. Yet you only see the ruinous results of race when your perspective of reality is affected.

For instance, when I appeared on television and radio, commentators frequently asked me about the "race card." They were invariably referring to Mark Fuhrman's testimony and the charges that he is a bigot who might have planted evidence to frame O. J. While Fuhrman may have been the ace of all race cards, the deck had been shuffled, and many hands dealt, long before his appearance for the prosecution.

For instance, the prosecution's and defense's choice of jurors featured denials of and concessions to race. Who would be more sympathetic to the defense, to the prosecution? Who favorably viewed the L.A. police department, who thought it stunk? And so on.

The choice to field Christopher Darden on the prosecution was a clear nod to race in this case. Darden seems oblivious to the fact that he was added to blacken up the prosecution's public face. His value derives not from his lawyerly demeanor or his rhetorical skills, which remain remarkably mediocre, but from his metaphysical presence in countering the incantatory powers of blackness invoked by Johnnie Cochran. The presence of Darden is meant to show that the prosecution is not racially insensitive. That O. J. and Cochran and Carl Douglas don't exhaust the resources of authentic black identity. The race card was played from the very beginning.

Marcia Clark is a brilliant attorney, a wonderfully disturbing presence. She is a woman with guts, whose fortitude and chutzpa are refreshing. The fact that she had to cut and restyle her hair, be harassed about the length of her skirts, and have her nearly nude picture plastered in a tabloid magazine underscores the hypocrisy of gender double standards. The men have wide latitude in dress; Marcia has to dress to please an amorphous constituency that she is partially responsible for cultivating. Who are these ideal jurors the prosecution played to in preparation for its case? Can they be trusted to reliably forecast her affect on twelve men and women already distressingly reshuffled because of internal disputes, inattentiveness, and claims of racial tensions?

It is odd to consider Marcia Clark as David to Johnnie Cochran's Goliath. After all, the state has enormous resources usually denied to the run-of-the-mill client that it prosecutes. Perhaps that's one reason people are pulling for O. J.: finally here's a guy who can literally afford to fight back. However, there are strident class divisions as well. Marcia

Clark and her colleagues are unglamorous, underpaid public prosecutors in battle with glamorous, highly paid private counsels. Clark is clearly the underdog, the woman who when she slugs it out with the guys on the defense is considered "whiny." Who when she stands up to the defense's shenanigans is considered aggressive. And who when she strategizes with cunning is considered disingenuous. (Think of her babysitting chores the night key defense witness Rosa Lopez was to give taped testimony. Spurred by the rancor caused by Cochran's rather rude, even sexist rebuke, Clark brilliantly, implicitly employed her domestic responsibility as a metaphor for how female identity had been unjustly hammered by the defense that night, and by extension, by their client O. J. Simpson throughout his entire relationship to Nicole Brown Simpson.)

Even Clark's attempt to identify with the undernamed victims, Nicole Brown Simpson and Ron Goldman, by wearing an angel on her lapel, was winning. To be sure, it showed poor judgment. But that's the quality that endears: a lawyer, a prosecutor no less, exposes the futility, the impossibility, even the undesirability, of objectivity by casting her lot with the deceased's families. Even though she lost her bid to keep the pin, Clark's gesture gave morality priority over legality.

Now that Clark's husband has waged a public war to wrest their children away from her, she has come to symbolize the ultimate contradiction of women working: if you don't work like a man, you'll lose your job. If you work like a man, you'll lose your family. Too often working women are punished by what are supposed to be their rewards. If Marcia Clark were a partner in a prestigious firm like Johnnie Cochran's, she could afford a nanny.

I have visited the crime scene. It is immediately the touchstone of my obsession with O. J. I view the impossibly small spot where Ron Goldman lost his life. I am able to even more vividly imagine the fierce struggle he waged to remain alive. And to see the landing on which Nicole breathed her last breath is numbing. I feel I have crossed some void, offended some greatly observed taboo against polluting privacy, even in death, especially in death. And yet strangely, I feel a tie to their lives. In the end, though, I realize that it is a bond produced by little more than a fetish for what television has made falsely familiar.

Why is it that some black people spend so much energy denying the impact of race on their lives, only to embrace it when their backs are against the wall? Clarence Thomas employed this ruse. He insisted that he had not relied upon race as a crutch to succeed. Still, when he was caught in the heat of battle with Anita Hill, he fell back on race. He used it in ways that only days before he would have disdained as cowardly and dishonest.

The same may be said about O. J. He always aspired to get beyond race, to be neither white nor black, to be human. That damnable equation has stumped many who have failed to understand that it sets up a false dichotomy. There is no such thing as being black and not already being a human being. The two are not diametrically opposed. Taken separately, at least for black folk, they are impossible.

When I read in O. J.'s book, *I Want To Tell You*, that as he faced racism in the past he either ignored or denied it, I am even more saddened. This sort of person is well known in many black circles. They protect themselves from racism by turning their backs on its most hateful expressions. They hide its most brutal effects in their spiritual or moral trunks. Or they cover the wounds racism inflicts with the temporary balm of diversion or avoidance. But sooner or later they must confront themselves and the choices they have made, particularly when the peace pacts they have negotiated with their psyches have disintegrated.

It used to be—or was it ever the case?—that the dead were safe, protected by their sleep beneath the ground. Nicole Brown Simpson and Ron Goldman have gained posthumous notoriety, for no other reason than their murders may have been committed by a fallen hero, a tarnished celebrity. It is the ultimate act of violation, a gesture of profound obscenity. Their deaths have been emptied of the inherently private meaning of grief to their circle of intimates and family. Their murders have given them a life beyond their bodies, but not a life respected by tabloid media or former friends looking to turn Nicole's and Ron's murders into money. The bottom line seems to be: it's not that they died, but that they were the persons that a famous man may have killed, that makes Nicole and Ron important.

It may be that football provided O. J. with a public context to wrestle the demon of violence that haunted his private life. The pure art of his movements on the field may have countered, driven, or even complemented his desperate scrambling for escape from inner turmoil. The ritualized cleansing of violent passions that brutal sports are alleged to achieve may only in the end lead to greater violence. The irony is that Nicole Brown Simpson benefited from the public face of O. J.'s acceptable aggression—a luxurious lifestyle as the wife of a sports star—but its private expression may have killed her.

Commentators have called the Simpson case the "trial of the century." What's intriguing about the case is how its major players are a virtual rainbow of color, gender, ethnicity, and class. Judge Lance Ito is Asian-American. Johnnie Cochran is

African-American. Marcia Clark is a white woman. And Robert Shapiro, like Clark, is Jewish. A judicial landmark is being constructed by people who a few decades ago couldn't stand equally together in the same court. The greatest contribution this trial may make to our country—besides the justice it may deliver for the murders of Nicole Brown Simpson and Ron Goldman—is to help ensure that representing the complex diversity of our nation becomes business-as-usual.

3

Gardner Taylor

The Poet Laureate
of the American Pulpit

"**G**ardner Taylor is the greatest preacher living, dead, or unborn," Wyatt Tee Walker proclaimed as he introduced Taylor in the fall of 1993 at a service marking Walker's twenty-fifth anniversary as pastor of Harlem's Canaan Baptist Church. (Walker gained fame while serving as one of Martin Luther King's trusted lieutenants.) Among black Baptists, the pastoral anniversary forms a distinct genre of religious appreciation. It is an often lavishly orchestrated event joining praise and pocketbook in feting a congregation's spiritual head.

But on the crisp October morning of his celebration, Walker shared the spotlight with the man

Time Magazine in 1980 dubbed "the Dean of the Nation's black preachers," a phrase that then New York Mayor David Dinkins would later repeat in his remarks at the service. After acknowledging Taylor's role as an adviser ("he used to tell me, 'Dave, you've got to bite bullets and butter biscuits'") Dinkins declared Taylor's preaching could be described in "only two ways: good and better."

These free flowing compliments might appear to be the natural excesses of a feel good service where the spirit is high and such praise, no matter how heartfelt, is by design the order of the day. But they mirror the sentiments of many more—black and white, religious and secular, preaching authorities and laypeople—who have been entranced, even transformed, by Taylor's legendary oratorical gifts.

Taylor, however, is more modest about his protean pulpit work. When I mentioned *Time*'s declaration, he deflected the tribute with characteristic humor. "You know what they say a Dean is, at least of eastern schools?" he asks, playing me with the instincts and timing of a seasoned comic. "Somebody too smart to be president, but not smart enough to teach." He smiled, shrugged his shoulders in self-deprecation and dead-panned, "So much for being Dean."

His humor and refreshing lack of hubris, combined with a preaching genius of extraordinary duration, have won the energetic seventy-seven-year-old Taylor a legion of admirers during his half century of ministry. Most of his career has been spent as pastor of Brooklyn's 14,000-member Concord Baptist Church of Christ. He made that pulpit perhaps the most prestigious in black Christendom before retiring in 1990 after forty-two years of service. The imposing, block-long gray brick church is a massive monument to black Christianity's continuing vitality in the midst of the well-documented decline of mainline religion. Under Taylor's leadership, Concord built a home for the aged, organized a fully accredited grade school (headed for over thirty years by Taylor's late wife, Laura), and developed the Christ Fund, a million dollar endowment for investing in the Brooklyn Community.

For Taylor, his success is an example of how God works in human life. "It is as if God said 'I'm going to take this unlikely person from the Deep South and I'm going to open opportunities for him to show [the world] what I can do,'" he says.

Taylor was born poor in Baton Rouge, Louisiana, in 1918, the only son of the Rev. Washington and Selina Taylor. "My father was a huge, tall, ebony man who had no trace of anything but Africa in him," Taylor says. "And he was extraordinarily arrogant about it." By contrast, his mother "looked white." After her husband's death, Selina Taylor attended "normal school" to become a teacher, later earning a degree from Southern University through extension courses. In one of his four books of sermons Taylor writes that despite his parents' lack of formal education, they "had a natural feel for the essential music of the English language wedded to an intimate and emotional affection for the great transactions of the Scriptures." The same is true of their son.

Although his father died before Taylor was thirteen, his father's influence, more than that of any other preacher's— especially his eloquent declamation and his wide range of reference—marks his son's preaching style. "Dad didn't finish high school, but he read voraciously. Sixty years ago, he spoke about Darwin's survival of the fittest and the battle of Thermopylae."

"Wash" Taylor enjoyed a wide reputation among Louisiana blacks for his brilliant preaching. Carl Stewart, Gardner Taylor's lifelong friend and a former basketball coach at Southern University, has for several years hosted a Baton Rouge radio show devoted exclusively to broadcasting the younger Taylor's sermons. Stewart illustrates Wash Taylor's preaching appeal by telling the story of a discussion between two Louisianans about an upcoming funeral. "'Hey, are you going to the funeral today?' one person asked. And his friend said,

'Who's dead?' And the other fella retorted, 'It really doesn't matter. Wash Taylor is preaching.'"

Despite his father's influence, Taylor attended Louisiana's Leland College in hopes of becoming a lawyer. "Clarence Darrow fascinated me," Taylor says in explaining his career choice. And because an aunt who helped raise him held the ministry in contempt, Taylor confesses that his view of religion wasn't exalted. "I didn't have the healthiest attitude about black preachers," Taylor says. "I thought preaching was a foolish way for people of normal intelligence to waste their lives."

But Taylor's plans changed dramatically when he survived a deadly automobile accident in which two white men died. Taylor experienced his "call" in that event, discerning God's claim on his life. "I thought that God must have wanted me to be his lawyer." Instead of enrolling at the University of Michigan law school where he had been admitted, Taylor ventured north to the now defunct Oberlin School of Theology. At Oberlin he read avidly, following writers ranging from Heywood Broun to Walter Lippmann. Their "literary styles affected me," he says. He also served as pastor of a church in nearby Elyria, Ohio, and after graduation, he pastored one in Baton Rouge, before being summoned at the tender age of thirty to Brooklyn's Concord Baptist Church, then with a membership of more than 5,000.

In New York Taylor joined an elite fellowship of ministers. "I don't think ever in the history of these two millennia have so many pulpit geniuses come together in one setting as I found in New York in the early '50s. . . . My God, it was unbelievable," he says. Taylor's multiracial aggregate included such preaching luminaries as Adam Clayton Powell, Jr., George Buttrick, Paul Scherer, Robert McCracken, Sandy Ray, and Fulton J. Sheen. Taylor has fond memories and wonderful stories about them all.

"Adam . . . with his angry oratory . . . was withering, blazing," he says of Powell, the controversial and colorful former pastor of Abyssinian Baptist Church and a longtime congressman. "He was a bonvivant. Adam had hair he could throw over his brow." Powell's nonkinky, straight mane along with his pale color, led to his being called "light, bright and almost white." "While we [black people] talk about the exaltation of our features, there was still [in the admiration of Powell's features] a lot left in us that adored white society," Taylor says.

"Buttrick [possessed] the poetry of the English Romantic poets. He was Wordsworth in the pulpit. He had a probing mind and relentless logic, and a gift for aphorism. For example, he said that the past ought to be a milestone, not a millstone." Paul Scherer, a former professor of homiletics at Union Theological Seminary, had a thespian bent. "Scherer was grand in manner. He had a great voice and a magnificent head of hair. As Jim Fry, a former student of his used to say, 'When Scherer said good morning, it was an occasion.'"

When Scherer was invited to deliver the prestigious Lyman Beecher Lectures at Yale University, "He said to a former classmate who was then a faculty member at Yale, 'You know, it's a great honor for them to have me here. I can't tell you how honored I feel. But why did they wait so long?'" Taylor relates, laughing at the story. Taylor delivered the 100th installment of the Beecher Lectures, which were published in 1977 as *How Shall They Preach*.

Taylor's unique blend of gifts may place him at the forefront of even this great cadre of preachers. His mastery of the technical aspects of preaching is remarkable. He brilliantly uses metaphor and has an uncanny sense of rhythmic timing put to dramatic but not crassly theatrical effect. He condenses profound biblical truths into elegantly memorable phrases. He makes keen use of parallels to layer and reinforce the purpose of his sermons. His stunning control of narrative flow seamlessly weaves his sermons together. His adroit mix and shift of cadences reflects the various dimensions of religious emotion. He superbly uses stories to illus-

trate profound intellectual truths and subtle repetition to
unify sermons. And his control of his resonant voice allows
him to pliantly whisper or prophetically thunder the truths of
the gospel. What was once alleged of southern Baptist
preacher Carlyle Marney may be equally said of Taylor: he
has a voice like God's—only deeper.

Taylor's commanding physical presence, hinged on a solid
six feet, one inch frame, suggests the regal bearing of pulpit
royalty. His broad face reveals seasoned character. His wide
set eyes are alive to the world around him. Taylor's forehead
is an artistic work of chiseled complexity. Furrows furiously
cross-hatch his bronze brow, extending to the receded areas
of his exposed, upper cranium where a shock of grey hair fas-
tidiously obeys its combed direction. Taylor's massive hands
are like finely etched soft leather. They function as dual
promonotories that stab the air in the broad sweep of pulpit
gesture or clasp each other in the steadied self-containment
of quiet reflection.

Taylor's snappy sartorial habits, though, hint more at Wall
Street executive than Baptist preacher. For a class on
homiletics that he occasionally teaches at Princeton Theological Seminary Taylor wore a dark blue, double-breasted wool
suit with a window pane design, a burgundy-striped shirt,
and a paisley tie. And at Wyatt Walker's pastoral anniversary,
he wore a charcoal grey pin-striped suit, with a white shirt
and burgundy tie.

But it is not his sharp dressing which draws most attention
to Taylor. The preacher's magnetism lies in his intimate and
unequalled command of the language and literature of the
English-speaking pulpit.

James Earl Massey, himself a noted preacher, professor of
homiletics and dean of Anderson School of Theology in Indi-
ana, ranks Taylor "as one of the top five unique pulpit
geniuses of any generation in American life." Massey contends that the gifts such figures as Harry Emerson Fosdick,
Phillips Brooks, and Henry Ward Beecher brought to the

American pulpit scene, "Taylor has brought in one person." Taylor possesses Beecher's "prolix ability to spin words," Brooks's "earnestness of style and breadth of learning," and Fosdick's "ability to appeal to the masses and yet maintain a dignity in doing so."

Preaching authority Henry H. Mitchell, author of the widely cited *Black Preaching*, points to Taylor's familiarity with the preaching tradition as a key to his appeal. "He's not only master of black preaching as such. He knows all the great white preachers and quotes them [as well]."

Carolyn Knight, a professor of preaching at Union Theological Seminary and highly regarded pastor of a New York Church, recalls a conversation she had with Taylor that displayed his endless pursuit of preaching excellence. "He told me last year—and he advised me to do it—that in preparation for his Beecher Lectures, he went down into the stacks of Union's library and read every set of published Beecher lectures."

Taylor's reputation as the "poet laureate of American Protestantism" is a considerable achievement. Throughout its history, black preaching has been widely viewed as a form of public address brimming with passion but lacking intellectual substance. Like black religion in general, black preaching is often seen as the cathartic expression of pent-up emotion, a verbal outpouring that supposedly compensates for low self-esteem or oppressed racial status. Not only are such stereotypes developed in ignorance of the variety of black preaching styles, but they don't take into account the black churches that boast a long history of educated clergy.

James Weldon Johnson's classic poem *God's Trombones* provides a literary glimpse of the art and imagination of the black folk preacher. C. L. Franklin's recorded sermons, spread out over sixty albums for Chess Records, brought the vigor and ecstasy of the black chanted sermon—dubbed in black church circles as "the whoop"—to the American public. By and large, however, Americans have remained insulated from the greatest rhetorical artists of the black pulpit.

Of course, broad segments of American society have sampled the richness of black preaching through the brilliant political oratory of Martin Luther King Jr. and Jesse Jackson. Both King's and Jackson's styles of public speech—their impassioned phrasing, intellectual acuity, and imaginative metaphors—reflect their roots in the black church. And their involvement in civil rights and politics extends the venerable tradition of black preachers serving as social critics and activists.

But their oratory—like that of preacher-politicians from Adam Clayton Powell to William Gray—has been shaped by the peculiar demands of public life and informed by a mission to translate the aspirations of black Americans to the larger secular society. The aims of their public speech have led them to emphasize certain elements of the black preaching and church tradition such as social justice, the institutional nature of sin, and the redistribution of wealth, while leaving aside such others as the cultivation of the spiritual life, the nurturing of church growth, and the development of pastoral theology. Such varying emphases are usually framed as the differences between "prophetic" and "priestly" religion. If the former has been most visible to American society in the guise of church based civil rights activists, the latter has been closer to the heart of the religious experience of most black Christians.

Though Taylor has combined both approaches—he was active in the civil rights movement in New York, and was a close friend and preaching idol of Martin Luther King Jr.— he realizes that his life work has ruled out the kind of visibility that comes from high profile activism. "I recognized early that the [kind of] work I do is not attention grabbing. . . . When I came along . . . college presidents were the lords of black America. Later, it became civil rights [leaders]. Still later, it became [holders of] political office."

Taylor humorously admits that with every attempt he made to "do something else [other than pastor], I got trapped." "They put the whole board of education off when I was a member," he says. "Governor Rockefeller wiped it out."

As James Massey maintains, "Taylor has stuck with the church. He has been busy handling the themes of the gospel, busy heralding what it was that Jesus came to do. He's been busy honoring the name of his Lord, shaping a community around that name and seeking to effect society in ways that are consonant with the gospel purpose. This is not newsworthy, like leading a sit-in."

Don Matthews of Washington's First Baptist Church, the church of fellow Baptists Bill Clinton and Al Gore, agrees. "We're in a time when the pulpit and the church in general are not particularly admired by anybody else that isn't in it. The only place it is perceived as powerful is in the political world. . . . But the people who have the spiritual word to speak aren't paid much attention by *The New York Times* or *The Washington Post.*"

William Augustus Jones, noted pastor of Brooklyn's Bethany Baptist Church, contends that Christians are "resident aliens" who have a radically different perspective on the world than secular citizens. "For a preacher to be regarded as popular means that his faithfulness to the word is not what it ought to be."

Ironically, Taylor's most bitter disappointments and defeats have come within the church world to which he has been single-mindedly devoted. Over dinner at Greenwich Village's popular Spanish restaurant El Charro's, Taylor, joined by his wife Laura—a shock of healthy black hair loosely pinned atop her hauntingly beautiful ebony face of high cheek bones and deep set eyes—recalls the 1952 fire that destroyed Concord Church.

"It was devastating. And were it not for this lady I don't know what I would have done," Taylor confesses. "She fooled me, innocently. The architect sat in our home, and . . . I asked him how much it was going to cost [to rebuild the church]. He said it would cost a million dollars. My heart went straight down. Black people in Bedford-Stuyvesant in 1952 couldn't raise a million dollars. But Laura said, 'Don't pay any attention to him. It won't cost a dime over $750,000.'" Taylor says that he was so anxious to believe her that for a year and a half he "traveled on that delusion." He laughs heartily as he reports that it cost nearly $2 million. In his view, his wife's figure turned out to be a "merciful deception."

Though naysayers said that Concord would never rebuild, the congregation not only erected an edifice on the very grounds of the fire but also added a gymnasium, an educational building, and a full space underneath the sanctuary, doubling the church's seating capacity.

Less satisfying was the outcome of the bitter 1960 confrontation between Taylor and J. H. Jackson, then president of the National Baptist Convention, USA, Inc., the nation's third largest Protestant denomination and the group to which most black Baptists belong. Jackson's conservative social and political views put him at odds with Martin Luther King Jr. and those ministers sympathetic to the cause of civil rights. Because of their disagreements about civil rights, and the issue of incumbency (Jackson had been president of the convention since 1953 and in the process broke convention limits on presidential tenure), Taylor agreed to run for the convention presidency at its annual meeting in Kansas in 1960.

A bitter fracas ensued. Hundreds of supporters of each candidate physically struggled and fought, leading to the accidental death of a loyal Jackson supporter and certain defeat for Taylor's team. The next year Taylor joined with King and other ministers who seceded from the convention to form the Progressive National Baptist Convention, Inc., which currently has a membership of more than two million people.

More than 30 years after this painful period, Taylor harbors no animosity toward Jackson, who died in 1991 at eighty-four. "Jackson had an ingenious and peculiar appeal to black people, as Reagan had to white people: J. H. Jackson could weep with the idea . . . that he was being put upon by powerful people . . . who were attacking [him], and he was weak. He had a gift for that."

Taylor even manages to find humor in illustrating this dimension of Jackson's appeal, which turned on his great gift of storytelling. H. H. Humes, a childhood friend of Jackson's recalls hearing Jackson preach about the hardships that afflicted his parents in attempting to send Jackson to college. "When it looked like he couldn't go back to college, his mother said, 'The boy must go.' And the father said, 'We don't have anything but a mule.' But she said, 'The boy must go to college.' And the father said, 'We won't have any way to get the crop.' But the mother said, 'The boy must go to college,' and they finally sold the mule. Humes was weeping as he came out of the church. And someone said to him, 'Well you grew up with him. What did they do for a crop [since] they sold the mule? [Taylor's voice affects a weepy tone] 'Oh,' Humes said, 'Jack's people never did own a mule but I just can't stand to hear him tell that story.'" Taylor breaks into laughter, trailing it with "Lord, have mer . . ." The last syllable of his plea is erased by more laughter.

Taylor's enormous gift of humor, his ability to acknowledge the humanity of his opponents, gives him the grace to accept and overcome his own failings. When I asked him about the thorny problem confronting the black church in its treatment of women in the ministry, Taylor confessed that he had to grow into his enlightened position.

"As with the white male, an exclusive preserve [of black male power is] under threat of invasion . . . I had to have a conversion myself. I knew theoretically this was wrong, but prejudice is not a rational thing."

Taylor's conversion occurred in the late '60s at Colgate Rochester Divinity School, where he was teaching a class on preaching. There Taylor encountered white female students whose fresh viewpoints helped change his views on women in ministry. "As they presented the gospel, I saw a new angle of vision . . . I had an interesting, and to me a humorous, thing happen. The young women of Colgate came to me and said, 'You know, you're just like all these other people here. You use all this sexist language.' Well, I was really stung. Who am I, having suffered from being excluded so long, to [exclude others]? And so I worked on it."

Taylor reports that after he delivered the Luccock Lectures at Yale Divinity School, a female faculty member thanked him for his inclusive language. Feeling good about the acknowledgment, he went back to Colgate to report this fresh triumph to his female students. "I saw some of the young women in the hall, and I told them what had happened. And I said, 'You know, I want to thank you girls . . . ' 'Oh,' they said, 'you don't say girls. Say women.' 'Well,' I said. 'It takes a little learning.'"

Since that time, Taylor has developed an acute analysis of gender relations, particularly in black culture. "I was greatly troubled by the Anita Hill–Clarence Thomas situation, and more troubled than almost anything by the attitude of [many] black women. I reached the conclusion that black women have been so put upon that they have developed a kind of psychological scar tissue, so that they've learned to take in stride things that ought to outrage them. And that distresses me. I am tired of the way black men misuse black women, and the way black women apologize for and accept what black men do."

It is above all Taylor's unsurpassed ability to preach to preachers—his keen sense of the preaching mission and its encumbrances and opportunities, its joyous peaks and its seemingly bottomless sorrows—that make him a popular presence among seminarians and seasoned preachers alike. In his homiletics class at Princeton, Taylor ranges through the history of the English pulpit with formidable ease, sharing stories of history's great divines. He whips out tattered pieces of newspaper, whose margins are covered with notes drawn from a massive and virtually infallible memory bank of preaching lore and legend.

On one of the days I attend his class, Taylor produces a snatch of paper ripped from the previous Sunday's *New York Times Book Review*, which he reads religiously, along with the *New Republic* ["I despise almost every word in it, but it gives me good targets to shoot at"], *The New York Review of Books*, and the daily *New York Times* and *Newsday*. Taylor reads a review describing the ingredients of a great novel—from its descriptive power to its presentation of a wide view of humanity without losing its link to individual characters. He reminds them that great preaching contains the same elements.

His desire to help other preachers has endeared Taylor to audiences across the nation. It has made his sermons to ministers legendary. Black preachers, especially, collect sermons with the zeal of avid fans of baseball cards. At conventions of black denominations, the tapes of famous ministers sell briskly. These tapes are especially circulated and reproduced among younger preachers, serving as models of preaching excellence and training in the high art of sacred speech. Some even preach the sermons to their own congregations, trying out fresh ideas and new words they have gleaned from master storytellers. Frederick Sampson's "Dwelling On the Outskirts of Devastation," Jeremiah Wright's "Prophets or Puppets," Charles Adams's "Sermons in Flesh," William Jones's "The Low Way Up," and Caesar Clark's "Elijah Is Us," have all acquired canonical status in a genre of religious address that treats the plight of the preacher. Several of Taylor's own sermons, including "Seeing Our Hurts With God's Eyes," and "A Wide Vision Through A Narrow Window," neither of which appear in his books of published sermons, are classics of the genre. They amply illustrate the astonishing range of his pulpit gifts.

In "A Wide Vision Through A Narrow Window," Taylor, speaking on a text from Job, details for his audience of preachers at Bishop College's L. K. Williams Institute in 1980, the price of authentic preaching. In an arresting metaphor he reminds us that for eighteen chapters Job's friends had turned against him, "driving cold steel into his already bleeding spirit." The nineteenth chapter—the chapter containing Taylor's text—is Job's "reply to their gloomy countenances, and their long, bitter indictment of his calamity, as they sit around the pallet of his misery."

Taylor employs and repeats the sermon's theme to sharpen his portrayal of Job's predicament, a condition where "the window has narrowed out of which he looks upon the landscape of life. Once there had been the homes of children, and fruitful fields, and lowing cattle and bleating sheep. But now, the window has narrowed." As if being forsaken by earthly friends were not enough, Job faces, as do all ministers, the prospect of feeling forsaken by God. Speaking through the voice of Job, Taylor says that "it seems that God has overthrown me. Now here is where the window *does* narrow to a slit. If God be for us, then what difference does it make, who is against us? . . . But, my father, if G-a-w-w-d be against us, what else is there left? I don't know why, in the solemn appointments of God, that there are times when it does indeed seem as if we've got not a friend in earth—or in *heaven!* "Taylor thunders. And then he emphatically completes the sentence, with a staccato verbal surge, "left!"

Taylor eloquently rephrases the theme of his sermon in question form. "And my brother preachers, you say that you want great power to move among men's heartstrings?" Taylor incants in almost mournful tones. "You cannot have that, without great sorrow. G-a-w-w-d can fill only the places that have been emptied of the joys of this life." He then challenges them with examples drawn from the lives of other suffering servants. "Dale of Carr's Lane in Birmingham [England] had particularly toward the end a terribly lonely existence. Charles Spurgeon in the Metropolitan Tabernacle [London] had rheumatism and gout that made life unbearable for him. Frederick W. Robertson of Brighton [England] was so sensitive that the least thing shattered him like the piercing of an eye. George Truett, who charmed the American South in the first four decades of this century lived in the after-memory of a hunting accident in which a friend was killed by his own gun."

Taylor restates his theme later in the sermon. He then implores his congregation of preachers to look beyond the peripheral signs of preaching greatness to the real source of pastoral insight—the common bond with one's hearers provided by suffering. "Now you may tickle people's fancies, but you will never preach to their hearts, until at some place, some solemn appointment has fallen upon your own life, and you have wept bitter tears, and gone to your own Gethsemane and climbed your own Calvary. That's where power is!"

Taylor rhythmically measures his speech, repeating the forces that cannot by themselves make for great preaching. He builds up tension for the ultimate release in the announcement of what constitutes the power of proclamation. "It is not in the tone of the voice. It is not in the eloquence of the preacher. It is not in the gracefulness of his gestures. It is not in the magnificence of his congregation. It is in a heart broken, and put together, by the eternal God!"

Taylor wrestles with some of the inevitable sadness that life brings—for instance, the suffering that comes with aging. "I have reached a very unflattering and unenviable time. I have more money than I have time. And that's not good. It was much better the other way. But then I'll take it—what can I do?"

As we discussed his fifty-two years of marriage to Laura, he said, "I sometimes see her lying in repose now, and [he pauses], a great sadness comes over me because I know one of us must leave the other. I don't know which I fear the most. [He pauses again]. But, what can we do?"

Tragically, Laura was struck and killed by a New York sanitation truck in February 1995. Despite this great loss, one suspects that Taylor will do what he has always done, whether life favored him or assaulted him. He will, as long as he is able, preach the Word of God. That has been his peculiar gift and burden, his bread and butter for more than fifty years. And out of his own suffering, he has shaped a ministry that has spoken to the hearts of men and women throughout the world. And because of his peculiar gift for making mortals see the light of God, no matter how dimmed by human frailty and failure, who can doubt that it will continue to shine on him in the hour of his greatest need.

4

Crossing Over Jordan

Let's start with the irrefutable: Michael Jordan is the sublime master of basketball. The man who led the Chicago Bulls to their third consecutive NBA championship in 1993, and who anchored the United States Dream Team at the 1992 Olympics in Barcelona is perhaps the world's most revered athlete. He remained a powerful icon even after transplanting himself from the courts to the baseball field. There he embarked on a minor league career that, while mediocre while it lasted, was nevertheless bracing in its nerve. It has been speculated that Jordan's return to basketball after a seventeen-month absence will effect American businesses to the tune of 2 billion dollars! Four games into his come back to basketball, Jordan put to rest any doubts about his continued mastery of the game. He netted 55 points against the New York Knicks, breaking his old record for the most points scored by an opponent at Madison Square Garden.

Unlike past sports heroes who have been constructed as demigods or political pawns, the thirty-two-year-old Jordan has never seemed beyond human. He wasn't born a genius on the court: He was cut from the varsity team his sophomore year in high school and put himself through his own practice drill during the summer to make it back on. He's had his less than savory moments: There was his widely publicized gambling troubles, his refusal to visit the White House after the Bulls' first championship season because he was already scheduled for a Nike endorsement. And then there was the publication of the book *Jordan Rules* by sportswriter Sam Smith, which purports to expose the seedy side of Jordan's heroic myth. His image remained untarnished. The fans still adored him.

Perhaps that's because Jordan's glory has always seemed to be particularly American; it starts with talent and ends with financial power. He has been propelled into a career that follows Bill Cosby's as the quintessential American pitchman. He and his skillful handlers have, with enormous calculation, cashed in on the Michael Jordan image and settled him into shrewd investments which will permit him to avoid the fate of older black athletes who lose their life earnings to piranhalike financial hangers-on. Jordan eats Wheaties, drives Chevrolets, wears Hanes, guzzles Gatorade, and consumes fast food at McDonald's.

And off Course, he sports Nikes.

Jordan may not have been the first basketball player to ink a sneaker contract, but he changed the art of the deal. Endorsements rarely had gone to team players in the past— instead golfers or tennis stars got the fat. But on the pure power of Jordan's popularity in the late '80's, he not only secured one of the largest merchandising contracts ever, but Nike sold over $200 million a year worth of Air Jordans during the five years of his basketball career before his initial retirement.

Though basketball is anchored in the metaphoric heart of African-American culture, Jordan has paradoxically transcended the negative meanings of race to become an icon of all-American athletic excellence. In Jordan, the black male body, still associated with menace outside of sports and entertainment, is made an object of white desire. And black desire finds in Jordan, through his athletic ability, the still almost exclusive entry into wealth and fame. He has become the supreme symbol of black cultural creativity in a society that is showing less and less tolerance for black youth whose support sustained his career. Jordan reflects black culture's love affair with spontaneity and improvisation, its brash experiments with performance, its fascination with those who exceed limits. Jordan's embodiment of these black cultural elements has created in American society a desire to, in the words of the famous Gatorade media campaign, "Be Like Mike."

In turn, Michael Jordan has helped the business world seize upon black cultural notions of cool, hip, and chic, which have undeniably influenced the look and sound of America. The deeper resonance of some of the styles can be problematic: the brand-name, high-style sneaker reigns as the universal icon of the consumer culture. It represents the distortion of social values, encouraging undisciplined acquisitiveness. Madison Avenue sends the message to acquire material goods at any cost, and that chant is piped into black urban centers where drugs and crime consequently flourish. Black youth learn to want to "live large," to emulate capitalism's excesses on their own turf. This force drives some black youth to rob or kill in order to realize their economic goals. The sneaker also reflects black urban realities embodied in rap culture and the underground economy of crack. (Many drug dealers are enamored of the high-style sneaker, and are often provided private showings by store owners.)

Jordan transcends these street wars. But no matter how high he soars, he still encounters racism. These incidents mock his desire to live beyond color—in his words, to be "neither black nor white," to be "viewed as a person." It's said, for example, that his reincarnation as a minor league baseball player came out of the thwarted passion he had for baseball as a youth. The unspoken apartheid in his home-town of Wilmington encouraged white kids to play baseball and black kids to shoot hoops. Even at the height of his fame in 1992, his father's murdered body languished for several weeks and was then cremated without being named, dubbed only "John Doe" by coroners who saw in the corpse of a black man just another unsolved case.

Jordan chafes under his indictment by black critics who claim he is not "black enough." But he has perhaps not clearly understood the differences between pursuing a life beyond the reach of racism and seeking one of racial neutrality.

The former is an attempt to deepen our understanding of how race—along with class, gender, geography, and sexual preference—shapes human experience. The latter is rooted in the belief that there could be a raceless being called "the person" who exists beyond not only the negative conse-quences of race, but also apart from the particular racial and cultural patterns that help shape human identity.

Despite the mixed messages they send, the cultural mean-ings that Jordan embodies represent a remarkable achieve-ment in American culture. We must not forget that a six foot six American of obvious African descent is the dominant presence in a sport that twenty years ago was belittled as a black man's game, unworthy of the massive attention it receives today. He remains a metaphor of mobility to the heights of excellence through genius and hard work.

5

The Soul of Sam Cooke

In 1954, two-and-a-half of the most sublime minutes of American song were recorded by a gospel group that received little play outside of its segregated world. Famous in church circles for their polyrhythms and tight a capella harmonies, the perfectly dubbed Soul Stirrers had already helped transform black religious music. But on the ballad "Any Day Now," the group's lead singer, Sam Cook (the "e" was added later), evokes a world teeming with cultural nuances hidden from white society while gesturing to the future of pop music. Though Cook is singing about going to heaven, he masks a complaint about earthly restrictions on black life by pining for a day when there's "no sorrow or sad-ness/ Just only complete gladness . . . ." But it's the way Cook bends the notes, shaping his desire for freedom in effortlessly undulated phrases, and the urbane emotion his performance revels in that hints at how this renegade gospel singer would shortly make pop history.

While Sam Cooke's artistic genius is widely hailed, it can't be adequately understood without knowing the world that produced him, that made his performance on "Any Day Now" a charged revelation. Admittedly, that's a daunting task. One has to grasp the ethos of black Pentecostalism, the effects of the Great Migration on urban black populations, the cultural impetus of the civil rights movement, the evolu-tion of black sacred music, the cultural habits of the Deep South, the history of American popular music, and the inti-mate rituals of black survival forged under oppression. But there's more.

One has to unpack the local histories of Chicago and Los Angeles, two very different black meccas in the early part of this century. Plus, one must have a comprehension of the technical aspects of black music and the cultural bulwark behind it that help us see how Cooke ingeniously employs the grammar of improvisation. No wonder biographers have shied away from Cooke's story. Getting it right demands the same sort of intellectual imagination and skill that marked Cooke's greatest art.

Fortunately, *You Send Me,* Daniel Wolff's exhaustive and eloquent biography of the fallen Black Orpheus meets the task superbly. Wolff lucidly depicts how Cooke's life molded his music, how his music changed his life. He highlights as well the gaggle of intriguing characters that surrounded Cooke, from the crotchety but creative Art Rupe, czar of Specialty Records, to Barbara Campbell, Cooke's second wife who remarried two months after he died. While this is certainly not a fawning tome—it chronicles the good and the bad, the edifying and the indecent alike—Wolff embraces his subject with critical appreciation.

Wolff starts his story where Cooke's life violently ends: with an Asian hooker named Lisa Boyer in a sleazy Los Angeles motel, shot to death by its black female manager, Bertha Franklin, on December 11, 1964, under circumstances that remain hazy and controversial. Franklin claimed she shot Cooke in self-defense after he assaulted her while seeking Boyer, who claimed she had rebuffed the singer and mistakenly fled with his clothes. Cooke's infamous demise tragically contrasted to his spiralling artistic ascent, and led most astute observers (including Wolff) to conclude that some measure of foul play surrounded the singer's death. But to those outside the field of Cooke's charisma—especially the white police who handled the case, oblivious to his stature—his slain body, as one of Cooke's friends suggested, was "'[a]nother nigger shot in Watts on a Saturday night.'"

Only thirty-three when he died, Cooke was poised to cross over to the sort of superstar status reserved for white artists like Frank Sinatra. Indeed, Wolff sees the furious contradictions that often bedeviled Cooke in his bid for the mainstream as the spur to artistic accomplishment. "His achievements were all about crossing over," Wolff writes, "whether it was passing through the restrictive covenants" of his rigidly segregated neighborhood in Chicago, "going from gospel to pop," or "making it at the Copa," the famous New York City nightclub that was a warm-up for Vegas.

In Wolff's view, Cooke failed to obtain pop deification (at least while he lived) in large part because racism denied him the wider audience he deserved. In many senses, he's right. While white teeny-boppers were condemned by their moralizing elders for listening to white rock 'n roll singers like Elvis Presley and Paul Anka, such disdain was no match for the lethal mix of fear and repulsion unleashed when Cooke's vocal magic entranced adoring white girls. Plus, white artists' covers of black hits—which for the most part failed to match the originals they shamelessly mimicked—gained them huge financial benefits denied to their black counterparts. Sadly, as Wolff makes clear, cultural appropriation and economic exploitation were the rule for most black artists.

Cooke combated both these forces ingeniously. In 1957, he started his own record company, SAR Records, with former manager J. W. Alexander, and with Soul Stirrer founder S. R. Crain (who, along with Clifton White and G. David Tenenbaum, aided Wolff with interviews and research). Cooke's entrepreneurial participation in the music *business* encouraged black artists after him to master their financial fates by controlling the lucrative publishing and production of their songs. Contemporary R&B songwriter, producer and artist Babyface— like Cooke flawlessly handsome, prodigiously gifted, prolific, and a domineering visionary with his artists—is unthinkable, perhaps impossible, without Cooke's pioneering efforts.

Cooke answered the appropriation of black musical styles by *re*appropriating them through his silky, sometimes saccharine, occasionally raw black pop aesthetic. What racism would restrict, Cooke would reinvent. His forays into pop genres from R&B to jazz, from Tin-Pan Alley to supper-club fare, allowed him to meet the world on his own terms by conceding *and* combating the dictates of race and the market. It was a tricky balancing act, one that led both to transcendent art and musical ephemera.

Cooke was familiar with such tensions. After all, he'd had to balance his religious upbringing with his worldly success. Though Wolff lays to rest an apocryphal story about Cooke being booed by a gospel audience after he crossed over to R&B, he deftly explores how Cooke's embrace of Saturday night secularism profoundly troubled his Sunday morning spirituality. Like Marvin Gaye and Donny Hathaway after him, Cooke's gospel roots proved both a blessing and a burden, a source of solace and scorn.

Still, among Cooke's most lasting contributions is that he helped bring the world of black gospel—itself the product of jazz, blues, and ragtime—to pop music, greatly influencing rhythm and blues and giving rise to soul and funk. Even Cooke's trademark melismas—emotionally embellished extensions of a syllable over several notes, especially his thrilling "whoa-o-a-o-o-o-o"—were drawn from a gospel world then little noticed by the mainstream.

Wolff brilliantly illumines that world. He gives a vibrant sense of the hidden charms, extraordinary achievements, tremendous competitiveness and moral contradictions of gospel's greatest performers. He casts light as well on the cultural exigencies and racial realities that shaped their art. And in brief scope he expertly portrays the Pentecostal theology of personal holiness that stamped gospel music and the denominational traditions its stars helped reinforce.

Cooke's father Charles, who as part of the Great Migration moved his family from Mississippi to Chicago in the early 1930s, was a Pentecostal minister. Wolff's evocation of the spiritual ethos of the Cooke home in Chicago, like that of thousands of other black families that belonged in the 1940s to the Holiness Movement (as Pentecostalism was termed), is dead-on. Wolff writes that the Holiness Movement was "conservative," offering its urban adherents "familiar, down-home values" that were "flexible enough to respond to the shock of what people were beginning to call the ghetto." Cooke and his five siblings revelled in the segregated world of Bronzeville, the eight-mile-long, mile-and-a-half-wide area of Chicago that "was the first major Negro urban center in the United States."

Even Cooke's first brush with celebrity wasn't enough to stem the tide of segregation. As a member of the famed Soul Stirrers, Cooke's considerable talents, like those of his colleagues, were rigidly restricted to an almost entirely black world. Cooke joined the Soul Stirrers in 1950 at the tender age of nineteen after singing with the local gospel group the Highway QC's. He replaced the legendary R. H. Harris, who had helped revolutionize gospel quartet singing with the double lead (two artists alternating in leading a song).

As Wolff notes, many deemed Harris "flat-out a better singer than Cook: a larger range, a sweeter tone, and a pioneering phrasing that Cook could only study." Indeed, Cooke began as a lesser imitation of the master. But he eventually found his voice when he transformed accidence into art. Straining during a program for a high note on a gospel number the Stirrers pitched to Harris's vocal range, Cooke slid deliciously under its high arch with a velvety, vibrant yodel— whoa-whoa-whoa—that delighted his hearers. The key to Cooke's discovery, and a trait that would distinguish his secular music, is that it was "an urban sound: cool, sophisticated, and yet shot through with emotion."

The Stirrers recorded for years on Art Rupe's Specialty Records, so named "because Rupe planned to specialize in citified Negro Music." The Los Angeles based independent label boasted a roster that included rock 'n' roll legends Little Richard, Larry Williams, and Lloyd Price. Tight-fisted and keenly entrepreneurial, and with an ear for black talent, Rupe exploited the musical gifts of the rapidly expanding "colored" citizenry of the 1940s, where a defense-industry boom increased Los Angeles' black population to half a million by 1943. But Rupe failed to see Cooke's crossover potential; he eventually had an ugly falling-out with Cooke over ownership of Cooke's secular songs when the singer hit big in the pop arena.

When Cooke crossed over to secular music, his smooth style, clear, sweet tone, and cataclysmic, combustive gospel undulations made him an instant star. "You Send Me" was his first big hit in 1957; his last, a posthumously released song that was a departure from Cooke's usual melodic pattern, was 1964's "A Change Is Gonna Come," a melancholy meditation on the state of race and a hopeful, insistent plea for transformation. In between, Cooke packed a heap of living and singing, and Wolff covers it, warts and all.

Wolff unblinkingly details Cooke's constant philandering during his two marriages, his out-of-wedlock children, his choice of career benefit over personal loyalty, his explosive temper. But Wolff also points out how Cooke exhibited extraordinary courage in combatting racism while performing. He makes clear, too, that even when he produced pop schlock, Cooke restlessly reached for the brass ring of cultural achievement beyond the artificial boundaries that race imposed.

And perhaps unintentionally, Wolff shows that Cooke's best work "in the world" was matched, perhaps surpassed, by the art he achieved in the more limited, but ironically, liberating world of gospel. There, Cooke's gladness and grief gained a hearing no secular arena could afford him. There, "Any Day Now"—or, for that matter, "A Change Is Gonna Come"—was immediately understood for the passionate, paradoxical mixture of protest and praise its lyrics and sound suggest. There, despite its crippling hypocrisy and its inuring sacrosanctities, Cooke's valiant attempt to confront the evils of the world might have been given a theology worthy of his struggle.

But the gospel world failed Cooke, too, by telling him that leaving gospel music meant leaving God. (Bumps Blackwell, who helped Cooke with his first secular hits, captures the moral contradictions of anti-pop sentiment in gospel as he declares preachers "say if you sing a pop song . . . you can't be religious. [Preachers] can be homosexuals; they can drink whiskey; they can do all kind of sinful things, but they can't sing a pop song"). The gospel world's confusion on this point, its poor doctrine of revelation, really, drove Cooke away, denying him the spiritual succor only it could supply. Thus Cooke's best art, in church and in clubs, was informed, even driven, by the pleasures and strengths each offered, even as it reflected the pains and troubles each inflicted.

As Wolff poignantly portrays, Cooke's gigantic gifts bridged worlds that had little contact, even less understanding of each other: the sacred and the secular, black and white, the chitlin' circuit, and the supper club. His earthy and ethereal performances are a monument to the extraordinary effort to speak in many tongues, and to reach beyond the limits of convention to tap the genius of unfettered self-expression.

6

The Lives of Black Men

Debates about black males in our nation have increased in proportion to black men's rising social misery. Their condition is indexed in withering numbers that barely detail the impact on their lives of informal drug economies, escalated unemployment, and stunning rates of homicide. The plight of most black males is so bad that some social commentators have come to refer forebodingly to black males as an "endangered species." Trapped between statistics and stereotypes, however, the brittle textures and uncomfortable truths of black male life are too often smoothed over to fit easily into pat explanations of either their prosperity or failure.

Brent Staples's poignant memoir, *Parallel Lives,* offers a glimpse behind and beyond the recitation of black male decline. The world in which Staples was shaped, a world where working class black people brought value and dignity to their labor and lives, is vividly realized. Staples is a writer of edifying eloquence whose hunger for memory—for the dirt and light of everyday experience as it is transformed by forces both natural and contrived—is fed by an obsessive will to observe.

He turns time and again in his memoir to the wombs that gave him life, from the community of black women in Chester, Pennsylvania, that gathered around his mother and granted him status as an "honorary woman" to the black male circles in which he ran, populated by brothers and cousins, schoolmates and neighborhood boys. Staples, who was born in the early '50s in the belly of an eastern industrial city, recalls the virtues of the work then available for black men. He writes, for instance, of their employment in the shipyards that dominated the area, or, as his father did, in driving trucks.

He returned years later to Chester after a decidedly unsettled childhood (the family moved constantly because of evictions), and to Roanoke, Virginia, where his family, sans father, had sought out earlier roots. There he confronts the postindustrial collapse of the job market, its harmful effects on black males no more powerfully illustrated than in the life and death of his own brother Blake, a drug dealer killed by a vengeful rival.

Blake's life and death provide the narrative bookends of Staples's remembrances. They are a mournful reminder that, increasingly, black male life is archetypically embodied in its violent, premature termination. With painful, pitiless honesty, Staples writes of his refusal to attend his younger brother's funeral, and how his repeated, failed attempts to rescue his brother from the inevitable consequences of his drug dealing exacted a costly moral toll. "I mourned Blake and buried him months before he died," Staples writes. "I would not suffer his death a second time."

At first blush, this seems a callous, cruel act of bad faith, the ritual sundering of family ties by the "successful brother" from a past too painful and embarrassing to bear. But as Staples's narrative makes clear, events take on a second life when memory and wisdom mingle to give purpose to the past. His decision to forego his brother's funeral became part of a life-long attempt to turn painful isolation into fruitful solitude.

Escape, sometimes thwarted, sometimes painfully realized, is a common motif of these memoirs. When Staples writes of attending college near his home in Chester—the only one of his family's nine children to do so—as a result of a chance meeting with one of its black professors, he underlines just how tenuously his escape was connected to chance, how luck often allows talent to shine.

A less successful attempt at escape—from the frozen regions of racial stereotyping—is symbolized in Staples's strange, patently quixotic stalking of Saul Bellow after Staples becomes a doctoral student in psychology at the University of Chicago. Staples jogged by Bellow's apartment and tried to force a chance sidewalk meeting because he wanted to *be* Bellow, to "steal the essence of him, to absorb it right into my bones." But Staples came away only with the harsh caricatures of black male life that he found in Bellow's fiction. He could not escape meaning far less to his idol than what Bellow meant to him.

Staples's frustrated quest expresses more than it intends; it symbolizes the attempt of black people to secure somehow the approval of a culture that deadens itself to the variety and complexity, the sheer ingenuity, of black life. But there is sometimes a little too much of the fascination with the exotic in Staples that blinds him to the allure of what is closer to home. It started early, too. He records that his friendships with black kids in his neighborhood were "largely disappointing" and that "The kids were just like me; there was nothing exotic about them. . . . The Polish and Ukrainian boys were alluringly exotic."

With *Parallel Lives,* Staples has surpassed the need for imitation, revealing a resolutely distinct voice as he negotiates the treacherous waters of racial identity in American culture. One need not agree with all of his conclusions about how those racial identities are shaped—much less how they should be politically employed—to appreciate this moving book. With its army of metaphors marching across the map of the unruly but rewarding terrain where personal experience reflects and refines national identity, *Parallel Lives* reminds us that the best personal writing is born of the courage to confront oneself.

7

Marion Barry and the
Politics of Redemption

Marion Barry's comeback victory, which made him Washington's mayor for the fourth time, is stunning but not surprising. Four years ago, Barry slouched toward political Armageddon with a prison sentence for smoking crack. But his supporters put him back in office as a councilman soon after he served his time. Why? Because Marion Barry tapped deeply into, and draws continuously from, an almost bottomless well of black resentment, victimization, and redemption.

Marion Barry found religion in jail. Nothing unusual there, especially for people in trouble. Charles Colson found God after thugging his way along with the Nixon junta; and the late Lee

Atwater confessed his sins almost in public as the shadow of death fell quickly and heavily upon his life. The difference with Barry, at least in the minds of his former constituents, is that the downfall for which he repented and found salvation was not his fault alone. How could that be? After all, Barry was captured in grotesque dishonor on tape pleading for sex, accepting, for him at least, another drug instead.

For many blacks, Barry's acquittal in the court of black conscience was a defiant prequel to Rodney King, an anticipatory rejoinder to white folk who, despite seeing King's body meet concrete and clubs with savage repetition, literally couldn't believe their eyes. For others, Barry's forgiveness among blacks was a desperate gesture of belief in figures who have caught white scorn in high places. When he put the pipe to his mouth, many reasoned, it was a vile if self-destructive confirmation of the establishment's will and power to do black leaders in with an assist from (Rasheeda Moore's) black hands. Thus, Marion the perpetrator, the clumsily effete dictator of an urban black colony silhouetted against white indifference, became Barry the victim.

Surely this logic is tortuous at a couple of turns. For one, Barry's D.C. had been in decline long before his ballyhooed tumble. Then too, Barry's peccadilloes precede by many years his entrapment: without them, the sting is void. Barry's victimization is made compelling only when the racial resentments that strangle D.C. are taken into account; powerful white folk who resented a flamboyant black mayor were resented by poor black folk themselves resented by powerful white folk. None of this was strictly color coded; there were black folk who were embarrassed by Barry and white folk who cut him slack. In the main, however, race ruled. Add religion to the mix, and the rest is both history and a glimpse of what's to come in Washington.

What Barry has managed to do is bridge the gap between spiritual rebirth and personal rehabilitation, between religious language and political desire. The ethical republic imagined by William James has been seized by the politics of conversion! Black leaders have historically drawn upon biblical stories of conquest by faith to power freedom movements through harsh periods of struggle. The increased secularization of black leadership—which under slavery and segregation sprung from the church, but under integration has steadily deserted the church's pews—has meant a distancing from the symbols of black religion. That change has also transformed the style of black politics. Outside of Jesse Jackson, Adam Clayton Powell's religiously embossed rhetoric has been largely whittled down.

Gone, too, for the most part, are theologically inspired interpretations of victimization. Black religious faith is at core an argument with evil, particularly racism, and an affirmation that black folk are able with God's help to turn bad occurrences to good effect. Black folk might be oppressed, but they aren't *mere* victims; they can overcome if they identify with a crucified God. Like him, they can shape their suffering into power, transform their victimization into moral victory. On this view, victims cease to be victims by exercising personal and social agency.

To be certain, the potential abuses of such an idea, when abstracted from its religious context, are legion. Many whites, especially, feel Barry fits this bill. Claims to victim status that lack the moral weight of real suffering to support them are indeed empty and harmful. Clarence Thomas's claim to be a victim of high-tech lynching comes easily to mind. And pimping the pain endured by one's group as an entitlement to do wrong, as many suspect of Barry, is itself a victimizing posture.

The real challenge for Barry as mayor is to remain true to the black religious tradition that by his own confession has rescued him. The way to achieve this goal is to practice the politics of redemption, to translate religious rhetoric about salvation into principled political practice. Barry's passionate acceptance speech on primary night evoked images of Bible-thumping revival preachers. His cadences took flight in nearly-sung passages ecstatically proclaiming his rebirth. That black folk in Washington are more willing than whites to give Barry another chance may speak to their charity and desperation. Or, perhaps, their culpable ignorance. Only time will tell.

LESSONS

politics of/and identity

8

Moral Panic or Civic Virtue?

When I graduated from college and won a fellowship to Princeton for work on a doctorate in religion, I was introduced to the passive side of the aggression some white men feel in face of the perceived gains of blacks. At the awards banquet where my fellowship was announced, a white man whom I had never met approached me.

"I wanna tell you something, son," he drawled in perfect Tennessee twang. "You're not going to Princeton *just* because you're black."

Well thanks for the tip, I thought, but I already knew that. I had made straight A's, in philosophy, my major field of study, had received an award as outstanding graduate in my department, and was among the top ten graduates in my class. But some uneasy emotion in this gentleman wouldn't let him rest until he unloaded onto me his presumptuousness about my intellectual achievements and his belief that my entrance into a prestigious university couldn't have happened without the aid of my color.

Writers like Shelby Steele are often exercised by such incidents. For him, this proves that white folk will forever think that black folk are getting a handout, that they advance in life with an artificial boost from affirmative action. For Steele, that stigma is simply too much to bear. The only way to get rid of it is to get rid of programs like affirmative action that unduly punish its black recipients by marking them as categorically inferior to whites.

The problem, of course, is that before affirmative action, whites often felt that their fellow black citizens were inferior. Except then, black folk were held in check by whites' perceptions of our inferiority because such perceptions were buttressed by law. Currently, black folk who are felt to be inferior can at least go on to achieve at high levels in arenas opened to them by affirmative action. Such programs merely make available to qualified minorities the jobs, schools, and resources that shouldn't have been unjustly shut away from them in the first place. If programs like affirmative action are the best way to undo past harm to blacks—and at present I think they are—then so be it. If black folk are stigmatized in the process, better to be stigmatized and treated justly than stigmatized without justice or power.

In many ways, Newt Gingrich's rise to power has come to symbolize the moral panic in our nation over programs ranging from affirmative action to immigration policies. American politics and public debate are now hostage to the most vicious form of "political correctness" yet to appear. If one rationally defends those with the most to lose in recent cultural and political battles—poor black women and youth, the working-class, brown immigrants, the fragile black middle class—one is written off as a paleoliberal or castigated as the prisoner of political naivete. The public spaces for dissent from moral panic and the new American mean-spiritedness are quickly shrinking. But that doesn't mean we shouldn't try to figure out creative ways to combat the often cruel attacks on those too poor, too powerless, and too persecuted to defend themselves.

We should begin by applauding the ingenuity but detesting the disingenuousness of those white males who have made themselves out to be the fall guys of affirmative action, the welfare state, and black progress. These white males—ranging from Gingrich to Wood and Custred, the two Berkeley professors behind the California initiative on affirmative action, and millions beside them—have perfected the politics of whining. They have appropriated the language of victimization they so violently opposed when deployed by blacks, women, and ethnic and sexual minorities.

But these white males have added a new, disturbing twist: they have distorted and at times erased history. Many white males, by underplaying the role of past racial or gender discrimination in shaping policies aimed at eradicating these blights, also deny the continued effects of racism and sexism on current social and economic arrangements. Sad, too, is the way white male paranoia thrives because of the historical vacuum that engulfs its most flagrant claims. For instance, many white males claim that preferential treatment for blacks under affirmative action has displaced employment and university slots for white males. But how do we know ahead of time that a certain number of school desks and jobs are meant (perhaps set aside?) for these white males? It is false to assume there was ever a specific, calculable relation between a "test score" and a job.

Building on this false assumption, many claim that merit alone is the deciding factor in the competition for social goods like education, employment, and the like. It is evident that such a notion is false when we recall that for years a percentage of alumni children were guaranteed automatic admission to several Ivy League universities. Merit is a contingent good, a politically determined quality whose use is governed by a range of considerations under specific conditions. It is therefore conceivable that race itself becomes a merit in a context where it has in the past been used to systematically exclude people from the benefits of our society. In fact, that's one of the major reasons affirmative action was developed. To overlook that history is to surrender present debates to historical amnesia.

But a further expression of moral panic and mean-spiritedness is glimpsed in the recent retaliation against brown immigrants in California's Proposition 187, which legally denies public aid, schooling, health care, and employment to illegal aliens. (I find it intriguing and ironic that 187 is also the Los Angeles police code for homicide made famous in gangsta rap lyrics; I suppose that piece of legislation certainly signifies the death of civic compassion in important ways.) This legislation aims social contempt indiscriminately at Spanish-speaking populations (after all, what's the difference between Mexicans and Puerto Ricans, between Cubans and Dominicans?). The legislation also unleashes resentment and bias on what will shortly be our nation's largest ethnic minority. Further, the legislation scapegoats people who themselves have been crushed by economic exploitation and cultural demonization.

Tragically, the antipathy to "them Mexicans" runs deep in many black communities. Unsurprisingly, black-brown tensions run highest where blacks and Latinos live and learn in close proximity. On a recent visit to speak at a west-side Chicago high school that was predominantly Puerto Rican with a small percentage of blacks, I witnessed predictable hostilities between the two groups. The differences involved culture, language, and the perception by black students that teachers and admininistrators cared very little for African-American life. When these differences are magnified in the larger society, they lead to even more lethal consequences.

For instance, cultural and racial tensions are revealed in rival gang activity focused on ethnic, racial and neighborhood turf. They also surface in ugly, divisive political disagreements between blacks and Latinos who reproduce the vicious stereotypes by which the other group has been victimized. Blacks and Latinos must resist such temptations and forge healthy coalitions built on realistic assessments of our relatively fragile places in American society. Both groups are under attack. Both should join hands when possible—around issues of unemployment and xenophobia, language and social stigmatization, and class and cultural prejudice. Neither group has the luxury of arguing over which is really more reasonable, more needy, or, in an ironic nod to the politics of moral panic, more American.

At root, the new American mean-spiritedness is really about redefining what a real American is. Newt Gingrich and many other right-wing white males have explicitly tried to un-Americanize anybody who disagrees with them, to paint them, as they tried with Bill Clinton, as the enemies of the common people. Such figures are hostile to a complex vision of American citizenship. Those who criticize Gingrich et al.'s vision of American conservatism as a mighty force undermining the best interests of working people and other minorities are often portrayed as disloyal Americans. Either one endorses their narrow views, or one is viewed as an antiquated citizen out of step with the "redeeming" purposes of the Republican revolution. By portraying himself and his cohort as the outsiders who have come to do battle with Washington insiders, Gingrich has shrewdly positioned himself as the defender of common people. Of course, his turned-back $4 million book advance, a $200,000 cable subsidy for a televised class, and his role as the second most powerful politician in our nation aside, Gingrich might have a case for being pretty common, if not average.

What all minority and progressive peoples should be about is aggressively asserting the *we* are Americans—not the only ones, not the most important ones, not the best ones, necessarily, but *real* Americans nonetheless. The mean-spiritedness and moral panic that has seized our body politic and clouded our public imagination must be challenged by the ethics of responsible representation of those Americans most hurt by such attacks. Only then will the best of our American heritage of civic virtue and civil justice be protected and extended. Those who callously deny these traditions may very well be responsible for compromising the integrity of the American character.

9

A Daughter's Pain,
a Father's Legacy

The allegation that Malcolm X's daughter Qubi-
lah Shabazz conspired to murder Louis Farrakhan focuses attention on two crucial issues: the
persistent role of conspiracy theories in
comptemporary black culture and the contradictory legacy of Malcolm X, assassinated in a
Harlem ballroom more than thirty years ago.

The Shabazz case brings to the surface a sub-
merged but powerful current of lingering suspicions toward the Government. It is well known
that for decades the Federal Bureau of Investiga-
tion carried out an aggressive campaign of moni-
toring and subverting black leaders and
organizations. From the Southern Christian

Leadership Conference to the Black Panthers, from Paul Robeson to Jesse Jackson, black organizations and activists were subjected to a range of perverse tactics designed to distort their public image and undercut their influence.

Such abuses dramatically increased the level of distrust, bordering on paranoia, of many blacks—not only toward the Government but toward each other, as fellow activists sometimes turned out to be informers. Since then conspiracies have continued to proliferate, a ragtag combination of fact, fantasy and fear.

Unfortunately, the proponents of such theories sometimes ignore real culprits lying closer to home. In the 1960s political groups like the Black Panthers and Nation of Islam had authoritarian, almost repressive tendencies; anyone who raised objections to the group's strategy or criticized its leaders was dealt with harshly. Conspiracy theorists were quick to blame outside manipulation for the resulting turmoil, when the source of the trouble was clearly within.

Many blacks were caught in a dilemma. If they acknowledged the vicious acts of intimidation by the Government, they were admitting that conspiracy theorists had a point. But if they played down Government interference, because they suspected black groups of exploiting conspiracy thinking, then they risked understating the extent of outside manipulation.

The Shabazz accusation dredges up many of these anxieties. Unquestionably, the timing of the Government's announcement provoked enormous skepticism. If it knew about the alleged murder plot for seven months, why wait until just weeks before the thirtieth anniversary of the assassination to bring charges?

It is probably fruitless to hope for an honest answer to this question. But more important, we should not allow this sad and bizzare turn of events to distract us from the more difficult business of reassessing Malcolm X's life and work. And such a reassessment is overdue, for it could point the way toward a moral and social climate in which it no longer makes sense to kill a human being because the vision he or she promotes is strange or threatening.

Perhaps the most harmful legacy of 1960s' militant nationalism is the idea that principles can be defended with violence. Though not always manifest in physical violence, intolerance is still expressed rhetorically in the black community. Malcolm X himself was known to engage in ad hominem attacks on other black leaders. But he was able to change his mind—primarily because he was also able to criticize himself, a quality that sometimes seems to be extinct among American leaders.

In one facet of his thought, however, Malcolm's views never wavered. His harsh indictment of the depredations encouraged in the ghetto—drugs, prostitution, hustling—reflected a fundamental puritanism. Despite his withering public assaults on white morality, Malcolm, like many blacks, breathed in the ethical atmosphere of mainstream America.

His conservative morality in a sense prevented him from truly respecting or understanding aspects of black culture. He saw mostly pathology in the conk, Lindy Hop, and zoot suit largely because he associated them with his past, the days when he was Malcolm Little, an unrepentant hustler. Yet like contemporary hip-hop culture—which Malcolm would most likely have condemned for its explicit sexuality and glamorization of drinking and drugs—black ghetto culture of the 1940s and 1950s must also be seen as a bulwark against the racism of the larger society that embodied the spirit of ordinary people.

Black youth today, for its part, has rightly fixed on Malcolm's rage at racism and economic misery. Angry words from his speeches have appeared in rap songs, on posters, and on T-shirts. One can only hope that his humanitarian legacy will eventually prove as powerful a resource. Surely this is the most lasting feature of Malcolm X's thought—the vision of social and personal transformation he expressed in his last year, when he renounced narrow nationalism and embraced people of all colors as brothers and sisters.

The thirtieth anniversary of his death reminds us that one of the greatest gifts black people can give to one another is the ability to disagree—and to turn our disagreements not into reprisals but into opportunities for communication and understanding.

10

The NAACP
and Black America

With the election of Myrlie Evers Williams as Chairman of its Board, the NAACP has made a strong gesture to recover some of its former glory. However the organization is still in crisis. The Ben Chavis conflict was merely a symptom of deeper trouble. Indeed, if the NAACP thinks it has solved its problems by electing Mrs. Williams and firing Chavis, it is avoiding a couple of tough lessons. First, it has lost touch with many African-Americans, especially the working class and the permanently poor. Second, it has strayed way too far from its radical roots.

The NAACP, like most civil rights groups, made racism its focus in combating the forces of inequality that strangled opportunities for blacks at the beginning of this century. Its specialty of attack was painstakingly shaped in our nation's courtrooms. Its pièce de résistance—the legal brief—was crafted by scholars and advocates like Charles Houston and Thurgood Marshall, who raised the defense of black interests to high, dramatic art. The results were sometimes revolutionary.

One was the monumental *Brown v. Board of Education*—the nation celebrated its fortieth anniversary in 1994—which attacked the destructive myth of separate but equal schools. With such cases to its credit, the NAACP's place in history is guaranteed.

Times have worsened for the black people the organization so nobly defended. Racism, certainly, continues to haunt and hurt black lives. Its lethal trace can be discerned despite subtle changes in the DNA of racial bigotry. But a specter hardly anticipated in the heyday of the civil rights movement has tragically materialized.

Sharp class divisions, which were masked and partially absorbed by the illusory uniformity imposed on black life under segregation, have ripped black communities apart. The gulf between have-gots and have-nots has only widened with the strategies the NAACP used to bring blacks into the economic mainstream. An unintended consequence of its success has been the trumpeting of differences *among* our nation's blacks even as it sought the social ground for common experiences *between* the races.

Until Chavis was hired, this fact seemed lost on the NAACP. Increasingly consumed with its corporate financing, the organization lost track of its less fortunate kin. The rank and file may have been working-class people, but its public face flashed the smile of smug bourgeois achievement.

Meanwhile, the poor, and particularly black youth, came under attack in political arguments that blamed their failures on their "pathological" culture, as seen in teen pregnancy, welfare dependency, urban violence, and rap music.

Like most black institutions limited by an exhausted liberal vocabulary, the NAACP had little energy or inclination to defend people whose experience was alien to its core constituency. For many poor and young blacks, it has become a relic, an ethnosaur.

Chavis's elevation promised a new day in the NAACP, with someone who would swell its membership rolls with the young and disaffected. But as he reached out to gang members, black urban youth, rappers, and controversial black leaders like Minister Louis Farrakhan, he curried wide disfavor with the NAACP's elite and mainstream black leadership. His critics argued that Chavis was aborting the organization's goals of mainstream acceptance and interracial cooperation.

Such logic reveals a huge non sequitur. Dialogue is not capitulation; conversation is not conversion. Only a cold war ethic of discourse could support such a notion. When black leaders speak to one another, even argue with one another, that doesn't mean they can't do the same with white allies. The idea that Chavis should not have met with or talked to certain kinds of black folk is a poisonous surrender to political correctness. After all, during the civil rights movement, black leaders often met with white figures who were explicitly, violently racist.

Even more troubling is the notion that to get along with the nonblack side of its constituency (read wealthy white and corporate donors), the association must steer clear of the vital issues that mold the lives of most black people. If this idea persists, the NAACP will stand for the national association for the advancement of corporate privilege, comfortable positions, and complacent policies.

Such a fate can be avoided if more black folks who can afford it will give to political and social organizations. It would then be less likely that groups like the NAACP would be straitjacketed, silenced, or bought off by interests inimical to black communities. Such patronage would bolster the prophetic and independent posture such organizations might adopt, especially if they are critical of corporations, big businesses, and market machinations.

Though many African-Americans are displeased with what happened to Chavis, how many of us have given money recently to the NAACP, the National Urban League, the National Rainbow Coalition, the Progressive National Baptist Convention, and similar groups?

Chavis may be guilty of poor administration and personal missteps, but he cannot be faulted for trying to change the organization's direction. Its guardians have it only half right when they claim that the NAACP has drifted from its base. They overlook the radical democracy and political progressivism of founding figures like W. E. B. DuBois. This pioneering historian and black cultural critic expressed with eloquence a complex vision of racial justice and economic equality.

The NAACP must recapture its radical dimensions precisely because the problems of black Americans—unemployment and urban violence, neoliberal rebuffs of racial progress and neoconservative assaults on young blacks, economic collapse and revived racism—are most assuredly not moderate. To insist that the NAACP needs to be its old moderate self once again contradicts its glorious, and radical, legacy.

11

Carol Moseley-Braun
and the Politics of Hope

If invention is the daughter of necessity, then improvisation is her twin. And political improvisation is something Carol Moseley-Braun knows a lot about. "I've . . . been able to deal with . . . politics as well as anybody, and I'm not surprised by any maneuver," Braun states.

In her stirring defeat of Illinois incumbent Alan Dixon in the U.S. Senate primary election in 1992, Braun's campaign drew from strong chords of feminist anger over Dixon's support of the Clarence Thomas nomination. Her campaign also benefitted from broad sweeps of black support in her bid to become the first African-American woman elected to the United States Senate.

Braun's feebly financed but feverishly fought campaign confounded the political pundits who, at best, looked upon her as a spoiler. Her victory sobered Chicago's political establishment and significantly changed the alignment of power in the Democratic Party in Illinois. "All I ask for is a level playing field," said a modest Braun, "because I believe I can best any of them."

Braun's political style—keen instincts and an ability to influence colleagues—recalls how her political mentor, Harold Washington, won Chicago's mayoralty. She copied his strategy of coalition politics, skillfully knitting together the interests of women, African-Americans, Latinos, and progressives to beat Dixon, who had not lost an election in his forty-three years in political office.

In clinching her victory, Braun gained 85 percent of the black vote in Chicago and Cook County. She also garnered 39 percent of the vote in suburban Cook County as well as collar counties that had traditionally been the stronghold of the moderate Dixon. Braun received 43 percent of the female vote, scoring heavily even among traditionally Republican female voters in a strong crossover vote. She received a surprising 34 percent of the male vote.

In November 1992, Braun faced Republican challenger Rich Williamson, a former Reagan aide and the beneficiary of a national Republican war chest that totaled almost $8 million. Braun initially sought to raise $3 million, and made several trips around the country to secure funds from party leaders and interest groups. She also charmed Mayor Richard Daley into pledging organizational support.

Braun's march toward a U.S. Senate dominated by rich white males began "in frustration," as she told *Today's Chicago Woman*. "[T]he system isn't working for us, and the people of Illinois haven't had the kind of leadership from our incumbent senator that we need and deserve," she explained.

Braun is no stranger to tough political fights. At forty-six, she has paid her dues as a politician. She served for three years as a prosecutor in the U.S. Attorney's office. She logged ten years in the Illinois House of Representatives where she was the first black and first woman to serve as assistant majority leader. From 1989 until her election to the Senate, she served as Cook County Recorder of Deeds. During her decade-long tenure as a legislator, Braun received for each year of service the Best Legislator Award bestowed by the Independent Voters of Illinois-Independent Precinct Organization (IVI-IPO). She also became known for her pioneering legislation in health care, welfare reform, education, South African divestiture, and minority and female set-asides.

Braun's bid for the Senate succeeded because she overcame the potential bane of all vigorous coalitional politics: tensions created by opposing interests within the fragile web of common political goals. While feminist backlash and black solidarity converged to boost Braun, such political cooperation might have as easily dissolved as differences over abortion sharply divided religiously conservative blacks from pro-choice women. Also, differences of belief in welfare reform, universal health care, and unemployment legislation introduced small but significant ideological wedges into Braun's broad base of support.

Braun's candidacy also tested the political resolve of black communities, which, like the mainstream, have traditionally shown strong evidence of gender bias. Braun's impressive showing in the primary among black voters promised a shift away from historic patterns of sexism within black communities. Such a shift will have to be substantially extended in order to reverse a lethal trend within the African-American political community. Braun's success in the November 1992 election was ensured as black political leadership aggressively pursued voter registration, political awareness, and community organizing among eligible voters who were disaffected from the political process. Such moves indicated a self-critical moment in black male political leadership and suggested a political maturity that bodes well not only for Braun's candidacy, but for the future of black politics at every level.

While pursuing her historic quest for national office, Braun did not alienate traditional Chicago Democratic party politicians. This enabled her to draw from the ample resources of the party establishment while articulating the claims of those constituents fundamentally disenchanted with politics as usual. Such is the highly skillful and deeply resourceful politics that Braun practices. "I make the distinction to people all the time between *government,* about which you have to be idealistic and have a vision and principles, and *politics,* which is the mechanics of getting there," Braun declares. "The reality is that politics is kind of rollerball. You never know what somebody's going to do, and you have to be prepared to deal with it."

As she integrates a U.S. Senate dominated by rich white males, Braun has the potential to express the grievances and wisdom of people who have been alienated, excluded or discouraged from participation in the political mainstream. Braun's election is the fruit of hard struggle in the trenches beside everyday folk seeking representation in the halls of power in an era of diminished political expectation. When she was growing up, Braun wanted to "be an adventurer, to set sail and explore uncharted territory and unknown lands." With her election, Braun has fulfilled her fantasies of worldly adventure. She has also achieved a pinnacle of political power and prestige that can give hope to other Carol Moseley-Brauns who may sometimes fear that politics is the art of the *impossible* instead of the possible.

12

King's Light,
Malcolm's Shadow

The renewed attention paid to Malcolm X's life
and career often trades on the assumption that he
was the polar opposite of the Rev. Dr. Martin
Luther King Jr. Malcolm was open to armed black
self-defense while King was nonviolent. True.
Malcolm blessed black rage by verbally releasing
it, while King preached the virtues of Christian
love. True.

Most forcefully, Malcolm spoke the uncompromising truth about black America while King, in
Malcolm's words, was a "twentieth-century Uncle
Tom"—the "best weapon that the white man" has
ever had. Nothing could be further from the truth.

Though Malcolm revised his views of King—

he later confessed that King's movement was more "militant" than he had initially believed—some members of a young black generation who claim Malcolm as an inspiration have failed to heed his change of heart. In lecture halls across America, I encounter young blacks who believe King was at best a moderate front for white interests.

Their perception that King was a sellout is rooted in two factors. First, it reflects the continued influence of certain varieties of 1960s black power and black nationalist rhetoric that often cast the debate about race in narrow, dualistic terms. Either one was for black separatist racial strategies or one was a pawn of white America. Sadly, this Manichaean perspective persists in pockets of black America.

Second, the belief that King undermined black interests grows out of ignorance about his intellectual growth. Awareness of his evolving thought has been smothered beneath white attempts to make him a safe American hero (witness the softening of his image in commemorative ads) and by attempts of some blacks to portray him as the tool of an amorphous white power structure. Neither view is true.

King began his career optimistic about the possibilities of erasing America's racial problems. He believed that fundamental change would be achieved by appealing to the consciences of white Americans. He was only partially correct.

At the start of the civil rights movement in the mid-'50s, America was bitterly divided by color and ruled by the practices of legal segregation. King's important voice in the civil rights struggle heralded a new day for blacks and whites by expressing a profound belief in the power of democratic principles to rid our nation of the plague of racism. He and his associates were largely successful, especially in the South where their efforts were concentrated. Public accommodations and transportation were desegregated, the Civil Rights Act was passed, the Voting Rights Bill was signed, and Southern apartheid appeared banished.

But the golden era of King's hope quickly passed, especially after its peak in the summer of 1963 when King forever etched into public memory the eloquent words of his Washington, D.C., oration, "I Have a Dream." After that summit of keen expectation, his dream dramatically receded, particularly with the Watts riots of 1965, when his doubts about the willingness of whites to work for a racially just society greatly increased.

As King grew more suspicious of our nation's ability to change, his language became more radical, his temperament less patient. Where before he had spoken of nonviolent civil disobedience, he began during the planning of the Poor People's March on Washington to speak about "aggressive nonviolence," and "nonviolent sabotage," by which he meant disrupting government and blocking roads during protests. By 1968, King said America was a "sick, neurotic nation" in need of a "revolution of values." He also became more critical of economic inequities, pointing out that America practiced "socialism for the rich and free enterprise for the poor."

Ten days before his death, King argued before the convention of the Rabbinical Assembly that "temporary segregation"—the maintenance of certain exclusively black schools and businesses, for example—may be necessary to prevent the loss of economic power that could result from complete integration. And in the last year of his life, King planned the Poor People's March, uniting poor blacks, whites, Latinos, and native Americans in a multiracial coalition that sought to challenge the unfair distribution of wealth, employment, and education.

King's later moral and economic principles put him on the outs with journalists and government officials, and he saw his popularity plummet before his assassination, at age thirty-nine, in Memphis on April 4, 1968. In a Jaunuary 1967 Gallup poll, King was no longer regarded as one of America's ten most admired men. Many of King's own aides were unwilling to commit their lives to the radical, fiscally Spartan lifestyle he adopted in observance of his mature views. Many whites disapproved of his changed views, and many blacks resented the broadening of his moral mission to include the waging of class struggle and opposition to war.

The columnist Carl Rowan called King's stance on Vietnam a "tragic decision," while Whitney Young, then head of the National Urban League, rebuked King for his position. King's scorching retort to Young was uncharacteristically blunt: "Whitney, what you're saying may get you a foundation grant, but it won't get you into the Kingdom of Truth."

Martin Luther King Jr. was a complex and flawed human being, a trumpet for moral and social revolution whose own failures are by now well known. No one who lived, fought and died as he did can ever be accused of selling out black people or compromising the principles of racial and economic justice. Like Malcolm X, King deserves a second look.

Race and the
Myth of Black Purity

One of the distinguishing marks of our nation is
that we are constantly torn between racial perceptions and racial practices. One major perception among many Americans is that not much
serious exchange takes place between races, especially between blacks and whites. The actual practices of race, however, suggest a complex pattern
of interrelationship between the races. Still, for
radically dissimilar reasons, members of dominant and minority cultures believe that race obeys
the rules of purity, that it is spared the very messi-
ness it calls into being. From what I have learned
about race in America, though, it is unruly and
impure. Its sometimes contradictory meanings

are as likely to upend the intentions of its vicious advocates as they are to find their fascist targets. In practice, race is far more complex than Americans perceive.

The notion that America has ever been *purely* anything, racial, sexual, religious, or otherwise, flies in the face of the heroic mixtures of all sorts that capture the breadth of American democracy. To say that America is composed of separate black and white nations, though, is a useful political fiction cobbled together from the fragments of historic black resistance to invisibility. It is meant to combat the unchallenged power of elites to name the state of affairs along the color line even as they exploit the belief that America is a largely realized dream. On this view, those who complain of its failure have only their recalcitrant black selves to blame. The appeal by blacks to two nations, or at least to a discrete black nation hedged by the relentless surge of utopian identity, is both a defensive and life-giving gesture.

For others, talk of two nations balloons past racial nationalism to land shakily on liberal premises or, worse, in the bog of dispiriting conservatism. For many liberals, economic and social disparities between blacks and whites prove the existence of two worlds governed by the unequal sharing of limited resources. For many conservatives, these same disparities illumine *moral* differences between the races and justify all manner of assault on black communities.

For nationalists, liberals and conservatives, the political usefulness of racial dualism obscures the cultural mixture among blacks, whites, Latinos, Asians, native Americans, and the like. Equally important, the force of class, the function of gender and the rise of gay sexuality to prominence in our culture chastens a strict adherence to race as the only lodestar of identity. These mixtures also challenge simple analyses of oppression that fail to grasp the creolized character of American identity.

In the meantime, racism appears to have outstripped its

theorists. Just when we thought that race had evolved from simple to complex forms—that myths of biological determinism had been largely cast aside in favor of the grittier, more promising realm of cultural incompatibility—the controversy over Murray and Hernstein's book, *The Bell Curve* arose. It's true that the old Manichaean ruse of "us" and "them" used to divide blacks and whites *racially* is often updated in a *culturally* determined "we" and "they." But as the popular appeal of *The Bell Curve* shows, such myths don't easily die. In either case, the bottom line is the same, to wit, no mixing or mingling. The vigorous attempts of some critics to replace biology with culture as the seat of social chaos, or to keep things as they are and give biology a prominent spot, encourages the nastiest sort of sentiments both here and abroad.

In light of the current state of race, conscientious Americans ought to attack myths of racial or cultural purity when and where they are found, whether in rosy cheeks or blackface. It would be silly to suggest, however, that even the most innocuous black nationalists have the moral or ideological backing enjoyed by subtle advocates of white cultural superiority. Still, the fetish of skin or culture betokened in breezy or romantic gestures of purity must be exposed and confronted.

The same impulse against purity must be extended to important debates about identity within black communities. (The same is true, by the way, for Asians, Latinos, native Americans, and other minorities.) Even as complex expressions of black genius have forced broad reconsideration of American identity, a defeating orthodoxy from within mocks the way black identity draws mightily from suppressed currents of sexual and gender difference. How can we imagine black culture without Zora Neale Hurston or Langston Hughes, James Baldwin or Mary McLeod Bethune? Furthermore, without the black gay passions of 1970s disco reheated in gangsta funk, a vibrant moment in black popular culture would be lost to boring bravado. Simply put, it's time for black folk to 'fess up, to admit that the quest for pure black identities is a sad imitation of the scourge of stereotype aimed at us from outside our hybrid existences.

We want to achieve a society where race doesn't rule, where race makes no difference that is tied to a form of authority that stigmatizes or punishes what falls outside its realm. The intent is not to erase race, but to nullify the harmful consequences grafted artificially to its expression. William James said that the difference that *makes* no difference *is* no difference. In the best of all possible worlds, race makes differences, but none that destroy the existence of other healthy differences.

Blacks and Jews on Stage

Anna Deavere Smith's play *Fires in the Mirror* is a haunting portrait of the small agreements and large grievances between blacks and Jews. Smith's characters telescope the complex rituals and values that hold racial communities together. Their resiliency is both tested and revealed during crises brought on by outside threats (in this case, the death of a black boy struck by a car driven by Hasidic Jews in Brooklyn, and the subsequent Crown Heights riots during which a Hasidic scholar was killed by a black youth). Smith's characters wrestle with racial calamity in broadly varying fashions. Some resort to intellectual defenses of religion while others rely on stylized racial politics. These strategies of cultural coping testify to how human identity is shaped by conflicting social forces.

Smith's pastiche of provocative personalities is postmodern. Her fusion of racial aspiration and anxiety in the condensed dialogue of selected characters from a real-life conflict is wrenching *théâtre vérité*. Smith's approach underscores and illumines the politics of identity, and the claims against it, that are presently sweeping college campuses across the nation. What Smith does exceedingly well is to show how identity politics are never negotiated with calm speech and rational action (though it often appears that way). Racial and cultural identity are constantly up for grabs even as they are ardently defended. Their outcomes are determined by the interplay between thought and practice, between reaction and reflection.

By tackling a specific racial tragedy—the death of an African-American boy and a Hasidic scholar—Smith highlights universal features of the turmoil between blacks and Jews, former compatriots and now estranged allies. Of course, things are different now than when Martin Luther King Jr. and Abraham Heschel joined hands to oppose oppression. Smith's play does not dramatize a black moral sensibility growing out of the civil rights movement belief in love at all costs. There is no talk of redemption through black suffering. The driving force behind many of the blacks she portrays is a chastened ethic of justice purchased by hard experience in the urban center.

Similarly, many of the Jews here are not familiar ones. These are not the Jews who have made their way into the mainstream of American culture by changing their names or practices to accommodate American fears of racial difference. So the lines of demarcation between blacks and Jews are rigidly and bitterly drawn in racial conflicts that expose the mutual ignorance about the other that persists in black and Jewish communities. Plus, the differences between blacks and Jews are often blurred in the haze of nostalgic talk

about a past zenith that has never been truthfully examined
or completely forged.

Smith's poignant vignettes are supremely important
because they explore racial difference without collapsing into
a paean about the virtues of our wonderful universality.
Smith forces us to seriously consider the particularities of our
conditions as people. She invites us to confront the racial,
social, and cultural events that bind us, even if loosely,
around a core of beliefs and behaviors that lend meaning to
our lives. Still, she reveals how such aspirations link us to a
transcendent human quest for stability in a rapidly changing
world. Avoiding a formal, didactic strategy that imposes
lessons from above, Smith allows the moral worldviews of
disparate communities to register in their own words—
sometimes harsh and bitter, but always honest and uncompromising.

Smith's play is an excellent resource for colleges and universities because it will evoke discussion about the broad
issues concerning the academy. These discussions include
identity politics as well as the idea of the relation of the particular to the universal. Her play is important for African-
American studies and American studies programs because it
forces students to think critically about the nature of race in
America. It forces us to think of how racial identity is composed of individual and social elements. This fact supplies
ample rebuke to those who would employ familiar mantras
and glib catch-phrases in place of hard thinking and fresh
analysis of our current racial predicaments.

The racial confusion on most campuses in a post–civil
rights era when the premises guiding social change are not
readily understood makes Smith's play an invaluable aid to
dialogue about race relations. Her play will be especially
helpful as our nation attempts to understand the furious
racial and ethnic catastrophes that have swept down throughout the world, from Bosnia-Herzegovina to Poland, from Ire-
land to South Africa. *Fires in the Mirror* confirms the

unfortunate accuracy of W. E. B. DuBois's 1903 prophecy that the problem of the twentieth century would be the problem of the color line. DuBois could have scarcely believed that his words would ring true for our next century as well. Smith's play, with all its passion and intelligence, is a small but powerful reminder that the force of race remains. But her work makes it clear that it is up to human beings, and the communities they make, to overcome the ignorance and fear upon which racism depends for its survival.

1 5

Portrait of the Black Family

Disraeli's famous quip that there are three kinds
of untruths—lies, damned lies, and statistics—has
never been more true than when applied to the
black family. For the past quarter century, from
Daniel P. Moynihan's 1965 report *The Negro
Family* to Bill Moyers's 1986 television special
The Vanishing Black Family, the black family has
come in for scrutiny by supposed "experts" wield-
ing a dizzying array of crunched numbers. The
point of all this data is often to prove contradictory cliches about the black family: it is either
going to hell in a handbasket or it is admirably
weathering the stormy social crises by which it is
continually buffeted.

For the most part, though, the common cultural perception of black families is indistinguishable from prevailing stereotypes: the sexual misbehavior of black teens, loafing welfare queens, black male neglect of domestic responsibility, and cultural pathology transmitted from one generation to another in patterns of destructive cultural activity. Andrew Billingsley's recent book, *Climbing Jacob's Ladder,* is a sharp and sophisticated rebuttal to the cultured despisers of the black family. It manages to accent the black family's strengths and successes while acknowledging its weaknesses and failures.

In a previous work written in 1968, *Black Families in White America,* Billingsley touted the virtues of black families in the face of withering indictments of that institution by commentators armed with Moynihan's report who claimed that the black family was falling apart. Billingsley argued that black families had survived the cruel injustices of American racism and shown a remarkable ability to create enabling patterns of familial stability and to maintain achievement in the midst of cultural chaos. Now, twenty-five years later, he supplies a richly detailed, lucidly argued, and statistically grounded elaboration of his earlier work. *Climbing Jacob's Ladder* greatly expands the scope and application of his thesis of the black family's survival under inhuman conditions.

The gist of Billingsley's argument is that "African-American families are both weak and strong but their strengths are by far more powerful and contain the seeds of their survival and rejuvenation." By mixing social theory, historical survey, and ethnographic analysis, Billingsley exhaustively details the social, economic, and cultural obstacles black families have overcome in preserving their racial and cultural efficacy. What is striking about Billingsley's portrait is the sheer *diversity* of black family life. In using that diversity as a theoretical point of departure, Billingsley erodes misconceptions about the homogeneity of black family life. He examines two-parent families, single-parent families, and no-parent families;

upper-income, middle-income, and low-income families; highly achieving, prominent families as well as socially marginalized and embattled families. All the while, he shows how slavery, racism, and poverty have exerted a deleterious effect on the structure and shape of black families.

Billingsley goes to great lengths to prove that black families adapt themselves to "the pressures and opportunities of the larger society in their constant effort to meet the needs of their members." Interracial marriages, for instance, reflect the changed racial climate and the adaptation of black families to this social fact in meeting the needs of various of its members.

Billingsley doesn't sidestep the familiar spiral of social forces that create crisis for black families. Teen pregnancy, criminal behavior, domestic violence, unemployment, and black male homicide are accounted for. Neither does he overlook the "generative elements" of black culture that make it "capable of providing resources and assistance." He explores, for instance, the role of the black church, the function of traditional black family values, and a powerful self-help tradition that has been prominent throughout African American history in helping "black families grow and move forward."

Billingsley's study will remind some, and inform others, that the black family can be no worse or better than the social forces that shape it and the opportunities that influence its development. In light of the failure to address the grave problems that assault black families from outside, Billingsley proves that it is an even greater injustice to merely blame black families for their problems within. That black culture has generated largely serviceable patterns of familial behavior is a testimony not simply to black culture's hunger for preservation, but to its creativity as well. Patterns of creativity and adaptation are revealed in "fictive kin" where black families are extended beyond the reach of blood or biology.

These patterns also shine through in single-female-headed households that are often heroic gestures of the will to survive. Billingsley's book helps to document that hunger and creativity, while offering a profound portrait of the complexity and outright ingenuity of black culture. These qualities are amply glimpsed through the beleaguered but brilliant vistas of black family life.

16

Screening the Black Panthers

In black communities, the politics of the 1960s
have resurfaced in the renewed popularity of
Malcolm X and black nationalism. The idealism
and rebellion of the '60s have been recycled in
song lyrics about the Black Panthers, and several
recent books have explored the history of the Panther Party. There has even been a revival of Afros
and black leather jackets. It seems inevitable,
then, that the Black Panthers, the most important
revolutionary black group to emerge in this century and one that brought style to politics, would
roar to life again on the screen.

Panther, a feature film by the director Mario
Van Peebles and his father, the writer Melvin Van

Peebles, tries to re-create the period of the Black Panthers' founding and to dress it in black-is-beautiful garb. In the movie, the time is 1966 through 1968 in Oakland, California, when Huey P. Newton (played by Marcus Chong, the nephew of Tommy Chong of Cheech and Chong), Bobby Seale (Courtney B. Vance), and a fictional character, Judge (Kadeem Hardison), form the Black Panthers and begin to recruit members.

In many ways *Panther* captures black nostalgia for the days before crack and crime gutted black neighborhoods, before the quest for upward mobility trampled the fires of revolution. It also has romance and, in a sign of how the Panthers have become fashionable again in black popular culture, is stocked with cameo appearances by black singers and personalities, among them Mark Curry, Kool Moe Dee, Dick Gregory, Bobby Brown, and the group Toni Toni Tone.

Although *Panther* creatively joins fiction to fact, anyone familiar with the period can see that the film is bold, revisionist history. The results are edifying and disturbing. The film accents some neglected truths while avoiding others, proving that cinematic imagination can be empowered or poorly served by a selective memory.

The portrayal of police violence against blacks is sledgehammered home appropriately enough. And the FBI (in the person of a dyspeptic J. Edgar Hoover, played by Richard Dysart) is accurately depicted as on an all-consuming mission to disrupt the work of black radicals.

But *Panther* is also willfully provocative, giving a dramatic spin to the facts, and it may bring former Panthers and others screaming from the woodwork. (An advertisement that ran in *Variety* on May 2, 1995, paid for by the conservative Center for the Study of Popular Culture, attacked *Panther* as a "two-hour lie," and said the Panthers "were cocaine-addicted gangsters who turned out their own women as prostitutes and committed hundreds of felonies.")

The film unabashedly favors the Panthers' perspective: it focuses on how they got started, what their goals were, and how they pursued them. In so doing, *Panther* adds the name of Mario Van Peebles to those of Oliver Stone and Spike Lee, directors who deny that history can be seen objectively. For these directors, film is inherently political; it's about taking sides and making points. While some of their movies bravely correct the fictions in accepted accounts of events, they also wear blinders of their own.

To be sure, *Panther* gets just right the heady mix of impetuousness and inspiration that drove Huey P. Newton and Bobby Seale to form the Black Panther Party in Oakland. *Panther* also vividly portrays the reign of terror by the white police in black communities. To those unfamiliar with the vicious activities of police officers in the '60s and '70s, the Van Peebleses' treatment will seem unnecessarily caricatured. It is, however, a necessary caricature, a gesture of cinematic exaggeration that faithfully evokes the spirit of police terror of that period. It was Rodney King made routine.

The triumph of the Black Panthers was that they symbolically thwarted the brutal police behavior in their community. When the Panthers patrolled the streets with their guns drawn—exploiting a law that permitted the brandishing of weapons—they struck terror in the hearts of many police officers and many whites, and awe in the hearts of many blacks. And as the film vividly shows, they used those guns, whether to shoot drug dealers or a police officer.

To paraphrase Pascal, however, the Panthers might have been guns, but they were *thinking* guns. They argued over how to implement the writings of Mao Zedong, Malcolm X, and Franz Fanon. *Panther* captures the intellectual ferment in the ghettoes of Oakland as Newton, Seale, and later Eldridge Cleaver (Anthony Griffith)—who joined the Panthers after writing the best-selling *Soul on Ice* in 1968—strategized against racism and capitalism. In the film's sustained attention to the Panthers' intellect, one can almost hear a plea for contemporary young blacks to hit the books.

When the Panthers rose to visibility, brandishing their weapons, it was believed that the greatest threat to the nation was a black man with a gun. (And unlike the white supremacist groups now rising on the right, the Panthers welcomed all races to the struggle.) They carried those rifles as a sign of rebellion and solidarity. The contrast with today's gang bangers who prey on the black community is there without being pressed home.

Even when the Panthers argue in this movie, they never approach the gratuitous violence and vulgarity of gangsta rap. Such a message hasn't been lost on the singers and hip-hop artists who appear on the soundtrack: such stars as Crystal Waters, TLC, Vanessa Williams, En Vogue, Mary J. Blige, Me'Shell Ndege'Ocello, Queen Latifah, Salt-N-Pepa, and SWV lift their voices together in "Freedom," the theme song from *Panther.*

Panther exhaustively explores the hidden history of the FBI's campaign to wipe out the group by encouraging internal dissension and through relentless surveillance. That, and much more, went on. But for the most part *Panther* ignores the intergroup violence that was only partly instigated by the Government.

The film barely touches on the internecine conflicts that ripped the Panthers into competing factions; nor does it broach the subject of the many members who were murdered by Panther loyalists. *Panther* pays scant attention to the heroic contributions of female Panthers like Elaine Brown, Newton's girlfriend, or Kathleen Cleaver, Cleaver's wife.

And the movie never addresses the sexual abuse, physical violence, and misogyny that were common in the Panthers' ranks, according to recent books by Brown (*A Taste of Power: A Black Woman's Story*) and Hugh Pearson (*Shadow of the Panther: Huey Newton and the Price of Black Power in*

America). By not alluding to the conflicts of gender, the Van Peebleses have robbed female Panthers of their due.

Panther taps deeply into the conspiracy vein to explain the Panthers' eventual demise. The film suggests that a deal was struck between organized crime and the Government to flood black communities with drugs in an effort to destroy urban black Americans along with the Panthers. To most people that theory is sheer folly. To some, however, it is a plausible answer to why heroin and then crack cocaine flooded into the ghetto, a subject Mario Van Peebles courageously addressed in his first film, *New Jack City*.

Perhaps both views have elements of truth. While there may be no conscious collaboration between political elites and drug lords to undermine black communities, the side effects of their deals have certainly harmed poor blacks. Like heroin in the '60s and '70s, crack flourished during the '80s in poor black communities, at the same time that the war on drugs failed to produce many victories.

In the end, drugs claimed the brilliant mind and battered body of at least one Panther, Huey Newton. Although he eventually received a Ph.D. in the history of consciousness at the University of California, Santa Cruz, in 1989 he was shot three times in the head during a drug deal that went bad. By that time Seale had turned to making barbecue sauce and Cleaver had become a neoconservative.

In the 1960s government repression along with internal dissension helped destroy the bulwark of the Black Panther party. The Panthers were neither thugs nor saints. They were soldiers of misfortune in a brutal battle against racist supremacy, vulgar capitalism, and the violent oppression of blacks. *Panther* helps us understand why this revolutionary Marxist group of the '60s armed itself.

SONGS OF CELEBRATION

17

Shakespeare
and Smokey Robinson

Revisiting the Culture Wars

I can hardly think of a subject more strained by confusion and bitterness than the relation of race to identity. Our anguish about this matter is at least three centuries older than the current turmoil stamped in the culture wars. American views on race and identity have wearily tracked our Faustian bargain with slavery, an accommodation of moral principle to material gain that has colored national history ever since.

The paradox of our situation is that Americans are continually fatigued and consumed by race. We sense, indeed fear, that its unavoidable presence is the truest key to our national identity. Yet we are as easily prone to deny that race has any but the most trivial affect on human affairs, and that it has little to do with personal achievement or failure. Therefore, the people whose lives have been shaped by the malicious meanings of race—to be sure, there are ennobling ones as well—must now endure the irony of its alleged disappearance in silence.

If they speak of the continued effect of racial bigotry, for instance, they are accused of exploiting unfairly their status as victims. If they talk of the injury inflicted by coded speech that avows neutrality even as it reinforces bias, they are called supporters of political correctness. If they appeal to black, or Latino, or native American heritage as a source of security in the face of hostility or neglect, they are said to practice the distorting politics of identity. And if they argue that Emerson be joined by, say Baldwin, in getting a fix on the pedigree of American literary invention—if they insist that the canon jams, occasionally backfires when stuffed with powerful material poorly placed—they are maligned for trading in a dangerous multicultural currency.

All of this makes clear that language is crucial to understanding, perhaps solving, though at other times even intensifying, the quandaries of identity that vex most blacks. Speaking and writing are not merely the record of our quest to conquer illiteracy or ignorance (they are not the same thing). Neither are they only meant to hedge against the probability of being forgotten in the future by marking our stay with eloquent parts of speech that add up to immortality. Language simply, supremely, reminds us that we exist at all.

Whether this is positive or negative, an uplifting or degrading experience, depends largely on how language—plus the politics it reflects and the power it extends—is used on our behalf or set against us. This is especially true for blacks. Early in American life the furious entanglements of ideology and commerce caused disputes about black folk to follow a viciously circular logic; slaves deprived of the mechanics of literacy for fear of their use in seeking liberation were judged inhuman and unintelligent because they could neither read nor write. Even those blacks who managed to show rhetorical or literary mastery were viewed as exceptional or hopelessly mediocre. However unfair, language became the most important battlefield upon which black identity was fought. This is no less true today.

The most important concerns of black life are intertwined in the politics of language—from the canon to gangsta rap, from the debate about welfare reform to the fracas about family values, from the roots of urban violence to the place of black religion. In my view a happy though unintended effect of the culture wars is that they force Americans to see that from the beginning our language has been indebted to political transaction.

It is not just now that ideological intimidation has allegedly ruined the prospect of objective judgment, or that its advocates have crashed the party and lowered the American standard of artistic achievement. Our literary traditions and rhetorical cultures eloquently testify to the influence of class upon taste, and reveal how power shapes the reception of art.

Black culture lives and dies by language. It thrives or slumps as its varied visions, and the means elected to pursue them, are carefully illumined or deliberately distorted. The threats, of course, are not entirely from the outside. The burden of complexity that rests at the heart of cultures across the black diaspora is often avoided in narrow visions of racial identity within black life. Its earnest proponents evoke the same old vocabulary of authenticity and cry of purity in their defense. But such moves echo as a hollow chant when voiced in league with the resounding complexity of identities expressed in the literature and music, the preaching and art of black culture.

Likewise, prolonged concentration on a fictitious, romantic black cultural purity obscures the virtues of complex black identity. An edifying impurity infuses black experiments with self-understanding and fires the urge to embrace and discard selves shaped in the liberty of radical improvisation. Fiction and jazz, for instance, urge us to savor the outer limits of our imagination as the sacred space of cultural identity. When advocates of particular versions of Afrocentrism and black nationalism claim a common uniqueness for black life, they deny the repertoire of difference that characterizes African cultures.

Such conflicts teach us to spell black culture and language in the plural, signifying the diversity that continually expands the circumference of black identities. If this is true for black culture, it is even more the case with American culture. The two are intimately joined, forged into a sometimes reluctant symbiosis that mocks the rigid lines of language and identity that set them apart. American culture is inconceivable without African-American life.

Can we imagine the high art of fusing religious rhetoric with secular complaint without Martin Luther King Jr. and Malcolm X? Their craft lifted freedom and democracy from their interment in ink and unleashed them as vital motives to social action. Can we think of contemporary American fiction, and its fiercely wrought negotiations with the cataclysmic forces of modernity, without the magisterial art of Morrison, Naylor, Wideman, and Walker?

Can we imagine the will to spontaneity, and what anthropologist Melville Herskovits termed the "deification of accident," that threads through American music without the artistry of Armstrong, Coltrane, and Ellington? And can we think intelligently about the American essay, that venerable form of address that splits the difference between opinion and art with felicitous abandon, without the elegiac anger of Baldwin and the knowing sophistication of Ellison?

These few examples point up the resolute dismissiveness that mark knee-jerk responses to multiculturalism at its best. Opponents to the opening of the American mind would have us believe that multiculturalism is the graffiti of inferior black art scrawled against the pure white walls of the American canon. This claim reveals how black cultural purists have nothing on the defenders of an equally mythic American literary tradition.

Among other influences, the American voice carries a British accent, even as it rallies to sublime expression the coarser popular elements of the times it both inhabits and transcends. It must be remembered that *Moby-Dick*, claimed by critics to be a work of Shakespearean magnitude by a writer of Shakespearean talent, gained such stature because Melville hitched the bard's cosmic grandeur to the motifs and genres of mid-nineteenth century popular literature. Imagine Hemingway doing a number on Jacqueline Susann, or Doctorow remaking Sidney Sheldon. The hybrid textures of the American grain are the most powerful argument for relinquishing beliefs in American orthodoxies and for celebrating the edifying impurity behind democratic experiments with culture and identity.

In this strict sense, multiculturalism doesn't argue for a future state of affairs to come into being. It simply seeks to bring to light the unacknowledged history of the trading back and forth along racial, and by extension, gender, class, and sexual lines. Multiculturalism is a request by minorities for this nation to come out of the closet, to own up to its rich and creolized practice in every corner of American life. In such an environment, it makes sense to ask, as Shelly Fisher Fishkin's poignant book about Twain's character does, "Was Huck Black?"

In a broader sense, though, multiculturalism cannot proceed painlessly. It must topple conventions precisely because they are erected on myths that exclude traditions and distort histories. The struggle over language and identity—over which work is legitimate and which is not, and over who gets to decide—is unmistakably a struggle of power. Plus, all the naysaying and hem-hawing that goes on around debates about multiculturalism neglect the manner in which great

African-American artists have often investigated both sides of the hyphen.

Ellison owed the habit of a critical style of reading, and the title of his first book of essays, to T. S. Eliot. Baldwin's essays draw equally from the gospel sensibilities and moral trajectory of the black sermon and the elegant expression of the King James Bible. And so on.

The fear of radical anti-multiculturalists that a democratized canon will trash Western tradition is mostly unfounded. At their best, multiculturalists expose the shifting contours of literary taste and the changing ways in which literacy is judged. (For instance, Homer could neither read nor write, but he is hardly frowned upon in our culture.) Multiculturalists also embrace the superior achievements of talented, towering figures. Such an operation bears little resemblance to hyperventilated protests that an ethic of racial compensation guides the selection of worthy work, and that its bad consequences will, in the words of Harold Bloom, "ruin the canon." I think of my own early education as an illustration of the possibility of black and white books together shaping a course of wide learning.

In the fifth grade I experienced a profound introduction to the life and literature of black people. Mrs. James was my teacher, a full-cheeked, honey brown skinned woman whose commitment to her students was remarkable. Mrs. James' sole mission was to bathe her students in the vast ocean of black intellectual and cultural life. She taught us to drink in the poetry of Paul Laurence Dunbar and Langston Hughes. In fact, I won my first contest of any sort when I received a prized blue ribbon for reciting Dunbar's "Little Brown Baby." I still get pleasure from reading Dunbar's vernacular vision:

> Little brown baby wif spa'klin' eyes,
> Who's pappy's darlin' an' who's pappy's chile?
> Who is it all de day nevah once tires

Fu' to be cross, er once loses dat smile?
Whah did you get dem teef? My, you's a scamp!
Whah did dat dimple come f'om in yo' chin?
Pappy do' know you—I b'lieves you's a tramp;
Mammy, dis hyeah's some ol' straggler got in!

Mrs. James also taught us to read Margaret Alexander Walker. I can still remember the thrill of listening to a chorus of fifth-grade black girls reciting, first in turn and then in unison, the verses to Walker's "For My People."

For the cramped bewildered years we went to school to learn to know the reasons why and the answers to and the people who and the places where and the days when, in memory of the bitter hours when we discovered we were black and poor and small and different and nobody cared and nobody wondered and nobody understood;

The girls' rhetorical staccatos and crescendos, their clear articulation and emotional expressiveness, were taught and encouraged by Mrs. James.

Mrs. James also opened to us the lore and legend of the black West long before it became stylish to do so. We read about the exploits of black cowboys like Deadwood Dick and Bill Pickett. We studied about great inventors like Jan Matzeliger, Garrett Morgan, and Granville T. Woods. The artists and inventors we learned about became for us more than mere names, more than dusty figures entombed in historical memory. Mrs. James helped bring the people we studied off the page and into our lives. She instructed us to paint their pictures, and to try our own hands at writing poetry and sharpening our own rhetorical skills. Mrs. James instilled in her students a pride of heritage and history that remains with me to this day.

Before it became popular, Mrs. James accented the multicultural nature of American culture by emphasizing the contributions of black folk who loved excellence and who passionately and intelligently celebrated the genius of black culture. She told us of the debates between W. E. B. DuBois and Booker T. Washington, and made us understand the crucial differences in their philosophical approaches to educating black people. There was never a hint that we could skate through school without studying hard. There was never a suggestion that the artistic and intellectual work we investigated was not open to criticism and interpretation. There was never even a whisper that the work we were doing was second-rate. There was no talk of easing standards or lowering our sights.

On the contrary, Mrs. James taught us that to really be black we would have to uphold the empowering intellectual and artistic traditions that we were being taught to understand and explore. Mrs. James was extraordinarily demanding, and insisted that our oral and written work aspire to a consistently high level of expression. And neither did she reproduce some of the old class biases that shaped black curricula around "high culture." She taught us the importance of Roland Hayes and Bessie Smith. She taught us to appreciate Marian Anderson and Mahalia Jackson. She encouraged us to revel in Paul Robeson and Louis Armstrong.

This last element of Mrs. James's pedagogy was particularly important since so many of her students lived in Detroit's inner city. She provided us a means of appreciating the popular culture that shaped our lives, as well as extending the quest for literacy by more traditional means. Thus, we never viewed The Temptations or Smokey Robinson as the raw antithesis to cultured life. We were taught to believe that the same musical genius that animated Scott Joplin lighted as well on Stevie Wonder. We saw no essential division between *I Know Why the Caged Bird Sings,* and "I Can't Get Next to You." Thus the postmodern came crashing in on me before I gained sight of it in Derrida and Foucault.

But Mrs. James's approach to teaching her students about black folk did no go over well with many of her black colleagues. Still bound to a radically traditionalist conception of elementary school curricula, many of Mrs. James's colleagues blasted her for wasting our time in learning ideas we could never apply, in grasping realities that would never give us skills to get good jobs. (This was still the late 1960s, and the full impact of the civil rights revolution had not yet trickled down to the classrooms, nor the psyches, of many black teachers.) But Mrs. James's outstanding example of intellectual industry and imagination has shaped my approach to education to this day.

Another event in my adolescence also shaped my quest for knowledge. I can vividly remember receiving a gift of *The Harvard Classics* by a generous neighbor, Mrs. Bennett, when I was in my early teens. Her husband, a staunch Republican (a fact which, despite my own politics, cautions against my wholesale reproach of the right), had recently died, and while first inclined to donate his collection to a local library, Mrs. Bennett gave them instead to a poor black boy who couldn't otherwise afford to own them. I was certainly the only boy on my block, and undoubtedly in my entire ghetto neighborhood, who simultaneously devoured Motown's music and Dana's *Two Years Before the Mast.*

I can barely describe my joy in owning Charles Eliot's monumental assembly of the "world's great literature" as I waded, and often, drowned, in the knowledge it offered. I memorized Tennyson's immortal closing lines from *Ulysses:*

Tho much is taken, much abides
And tho we are not now
That strength which in old days moved earth and heaven
That which we are, we are
One equal temper of heroic hearts made weak by time
    and fate but strong in will
To strive, to seek, to find, but not to yield.

I cherished as well the sad beauty of Thomas Gray's poem "Elegy (Written in a Country Churchyard)," reading into one of its stanzas the expression of unrealized promise for black children in my native Detroit:

> Full many a gem of purest ray serene
> The dark unfathom'd caves of ocean bear:
> Full many a flower is born to blush unseen
> And waste its sweetness on the desert air.

I pored over Benjamin Franklin's *Autobiography* and exulted in Marcus Aurelius; I drank in Milton's prose and followed Bunyan's *Pilgrim's Progress*. I read John Stuart Mill's political philosophy and read enthusiastically Carlyle's essays (in part, I confess, because his quote, "No lie can live forever" had become branded on my brain from repeated listening to Martin Luther King Jr.'s recorded speeches). I read Lincoln, Hobbes, and Plutarch; the metaphysical poets; and Elizabethan drama. (This last indulgence led a reviewer of my first book to chide me for resorting to Victorian phrases—which in his view was patently inauthentic—to describe a painful incident of racism in my life; I was tempted to write him and explain the origin of my faulty adaptation, but, alas, I concluded that "that way lies tears".)

The Harvard Classics whetted my appetite for more learning, and I was delighted to discover that it opened an exciting world to me, a world beyond the buzz of bullets and the whiplash of urban violence. One day, however, that learning led me right to the den of danger. Inspired by reading the English translation of Sartre's autobiography *Les Mots* (*The Words*) I rushed to the corner store to buy a cigar, thinking that its exotic odor would provide a whiff of the Parisian cafe life where the aging master had hammered out his existential creed on the left bank.

My fourteen-year-old mind was reeling with anticipation as I approached the counter to confidently ask for a stogie. Just then, I felt a jolt in my back; it was the barrel of a sawed-off shotgun, and its owner ordered me and the other customers to find the floor as he and his partners robbed the store. Luckily, we survived the six guns brandished that day to take our money. Long before Marx and Gramsci would remind me, I understood that consciousness is shaped by the material realm, that learning takes place in a world of trouble.

I was later thrilled to know that the new pastor of my church, Frederick Sampson, a Shakespearean figure if there ever was one, shared my love of learning. An erudite man trained to speak the King's English to the Queen's taste, he would, at a moment's notice, embellish his sermons and conversation with long stretches of Shakespeare or Wordsworth. Even at funerals, as he led the procession out of the church, he would recite:

> Life is earnest, life is real
> and the grave is not its goal
> Dust Thou Art, to dust thou returnest
> Was not spoken of the soul
> Let us then be up and doing
> With a heart for any fate
> Still pursuing, still achieving
> Learn to labor and to wait.

But like Mrs. James, Dr. Sampson read widely in black letters. Time and again, his eloquent pulpit art indexed the joys and frustrations of black and religious identity. He ranged between unlikely sources to make his points. He called on Bertrand Russell ("the center of me is a wild curious pain . . . [the search for God] is like passionate love for a ghost") and W. E. B. DuBois:

> It is a peculiar sensation, this double-consciousness, this sense of always looking at one's self through the eyes of others, of measuring one's soul by the tape of a world that looks on in amused contempt and pity. One ever feels his twoness,—an American, a Negro; two souls, two thoughts, two unreconciled strivings; two warring ideals in one dark body, whose dogged strength alone keeps it from being torn asunder.

Mrs. James, Dr. Sampson, and my early habits of reading are to me models of how the American canon can be made broad and deep enough to accommodate the complex meanings of American identity. To embrace Shakespeare, we need not malign DuBois. To explore black identity, we need not forsake the learning of the majority culture. And even if Dostoyevsky never appears among the pygmies, great culture may nonetheless be produced in unexpected spots.

The difficulties of gaining clarity about cultural and racial identity are only increased with the introduction of theory into the mix, a move bitterly debated among black intellectuals. The application of theory to black culture has provoked resistance from the right and left alike, mimicking patterns of response to theory in larger literate culture. Now what is meant by theory is literary theory, not a theory of progressive politics, say, or a theory of quantum mechanics, though both have come under attack for sharply different reasons.

The notion of theory itself, however, is not suspect. How could it be? Even its opponents have theories about the problems with theory. Some Marxists and feminists have theories about why deconstructionists need to be more realistically grounded in the world and politically engaged. Defenders of the Great Books have ideas about why theorists romp in pedantry and obfuscation, their jargon a sign of poor writing, or worse yet, muddled thinking. African-Americanists have theories of why black intellectuals should spurn European theories and stick to more traditional ways of criticizing books and culture. In their opposition to theory, at least, usual opponents find full agreement. With some adjustments, I think theory may help to explain black culture. We must have at least two skills to make it a go.

The first skill is *translation*. What's said meaningfully in one place must often be restated to make sense in another setting. Among initiates, subtleties of theory will be transparent, while those outside the theoretical loop will inevitably miss out. But if theory is to serve or undermine traditions of interpreting books and culture, the moral of the story (even if the point is that there isn't one) must at crucial points become clear. Admittedly, that is sometimes communicated by writers whose politics of expression lash out at simple, given meaning. In order to be successful, though, such an act should not be hindered by sloppy execution. As with all writing, there are good and bad ways to do theory.

The second skill is *baptism*. I know the phrase evokes volatile responses because of its religious association, but then I've got a theory or two about that. For Lyotard, Derrida, and Foucault to be useful to me, they can't be dragged whole-hog into black intellectual debates without getting dipped in the waters of African-American culture. Strategies of play, notions of "difference," and ideas about the relation of knowledge to power can illumine aspects of black culture when applied judiciously.

But theory must be reborn in the particular cultural forms that shape its use; it must reflect the cultural figures fixed in its gaze. Jazz and science fiction, hip-hop culture and collagist painting, and broad intellectual imagination—embodied in folk like Betty Carter and Octavia Butler, Snoop Doggy Dogg, and Romare Bearden, C. L. R. James and Zora Neale Hurston—all have something to gain from, and to give to, theory. Bertrand Russell believed that the goal of education is to help us resist the seductions of eloquence. At its best—in translation and baptized—theory can do just that.

The controversies surrounding hip-hop bring us full circle in grappling with how race, language and identity are joined, and how their contradictory meanings sometimes collide. Because of its extraordinary visibility, indeed, vilification in the larger society—and because of the strong veto it has aroused in many black quarters as well—rap perfectly symbolizes the failure of neat, pure analysis to illumine the complex workings of black culture. The debates about hip-hop culture strike the deepest nerves in black culture—how we name ourselves; how the white world views us; how we shape images and identities that are tied to commerce and exploitation; how black culture preserves itself while continually evolving; and finally, and perhaps, most important, how survival is linked to the way words are used for and against us. Like the black culture that produces it, rap is both a new thing, and the same ol' same ol'. That is the crux of black culture's gift and burden.

As debates about the canon continue, and as currents of suspicion about the wisdom of multiculturalism endlessly swirl, the example of black culture's constant evolution and relentless self-recreation is heartening. At its best, African-American culture provides an empowering model of education that combines the impetus to broad learning and experimentation with new forms of cultural expression. The ongoing controversies generated by identity politics, hip-hop culture, and racial politics, and the insurgence of a host of other minority voices, insures that African-American intellectual and cultural life remains an important resource in addressing not only marginal traditions, but in reconceiving and expanding the very framework of American literature and democracy.

Minstrelsy or Ministry?

One of the more suspect items to find its way into my mailbox recently is a new transliteration of the Bible; this time, through the use of over-wrought and tendentious slang, the Good Book has been aimed at black youth in the vain hope they will be drawn back to church if it speaks to them "in their own language" (or someone's vague understanding of it).

Making the sometimes impenetrable, often alien language of the Bible come alive in lucid prose is a longstanding goal of Christianity. The Greek text from which many biblical translations of the New Testament are derived was written not in classical but Koine Greek, the language of the common citizen. When it was introduced, for example, the King James Version caused quite a stir. In many learned quarters, it was deemed a heresy to compromise the beauty and purity of scripture by translating it into what was then everyday English.

Later still, advocates of the Revised Standard Version (dubbed in some religious circles the *Reviled* Version), the Goodspeed Translation, and The New English Bible came under attack in religious circles for departing from the revelation of the revered King James Version. Yesterday's controversial translation is often today's Biblical treasure.

Transliteration is another thing altogether. Instead of relying upon the original Greek text and translating it into the common language of the day, transliterators turn to existing translations to render their versions of the Word. Transliterators take more liberties, primarily because they lack a sense of the verbal nuances and constraints of interpretation that intimate familiarity with Biblical languages and customs might impose. The most successful transliteration to date is The Living Bible, a popular Bible employed by millions of Christians to, as the saying goes in the Black Church, "make it plain."

Now comes the *Black Bible Chronicles,* by Patricia McCary, purportedly to bring the Bible to the street in the slang of black culture. The famous serpent of Genesis is "one bad dude." God's creative competence means "the Almighty knew His stuff." And before the earth was formed it "was a fashion misfit, being so uncool and dark. . . ." For this to be what author McCary has called an "Afrocentric" Bible, equating uncool and dark rings false. And obviously, the Chronicles retain the sexism of older versions. These are merely little tips that Chronicles is plagued. From its packaging to its presentation and presumptions, this Bible is chock full of trouble.

First, the advertisement accompanying the book, a transliteration of the first five books of the Bible (known as the Pentateuch) reads: "Controversial? Sure. Which is just one of the reasons why this will be a sure seller." So much for the holy mission of winning souls and saving lives! Even when confronting the necessity of marketing strategies that selling any book presents, that's pretty crass fare.

Next, there's a foreword by Andrew Young. The minister and civil rights activist touts the *Chronicles'* virtue by comparing its production to the work of "Martin Luther and others," who "translated the Bible into the language of the people of their day." Young also contends that the *Chronicles* will bring the Word "to our younger generation in contemporary language." Say what? The appeal to authority, a fallacy taught in all basic philosophy courses, is raging at full throttle. First, one has to wonder whether Andrew Young knows a whole lot about contemporary black youth—about the ways they communicate in complex cultural rituals and language that takes more than being black to understand. Screaming from his words, and the *Chronicles'* backers, is a noblesse oblige derived from a caricature of black youth culture that is quite disturbing.

Nobody should know more than Rev. Young that in black churches across this country from storefronts to standing room-only sanctuaries of thousands—the quest for literacy is a prime passion of black folk, especially black youth. And even among those black youth to whom this book is presumably addressed, those black youth who give a hoot about the church or religion, the quest for self-improvement and survival has taught them about the need to be multilingual. They must learn the codes of street speech, of common English usage, and of the particular dialects and social accents of their own region or turf. *Black Bible Chronicles* assumes that slang is constant and unvarying, that black folk are not different from one another, that East and West Coast, for instance, are the same. (Do they call a brick an "alley-apple" in Denver?).

There is a desperate need to reach young blacks, to touch

them where they live. But stereotypes about black youth culture can't replace the hard work it takes to get inside that culture. Indiscriminate, anachronistic slang can't do the job that real love and respect will to win their appreciation and gain their ears and eyes. The real challenge is to translate the gospel into action—to sever it from bourgeois respectability and moralistic condemnation and to use it as a weapon of the suffering and oppressed to make justice a reality. What black youth need are jobs and spiritual renewal, moral passion and enough money to act with decency and self-respect. Then, they can afford to be themselves in a world that respects the variety of black life. Then they can heed the words of an elderly woman in the Black Church: "Be who you is and not who you ain't; cause if you is what you ain't, you am what you not."

19

Everyday Is Sunday

Like the classic era of American film, gospel music's golden age is in the past. Though contemporary stars like BeBe and CeCe Winans and Take 6 shine in the firmament of musical acclaim, gospel's greatest pioneers hail from a creative period that achieved its height during the 1960s. Though a few garnered wide notice (and most could have gained wider recognition still had they forsaken gospel and sung secular songs) they largely labored in anonymity, except to the faithful who viewed them as the geniuses their talent—and considerable egos—proclaimed.

Still, gospel's influence is felt deeply in the history of rock, pop, and soul—in the compressed frenzy of Marion Williams's ecstatically ejaculated *whoo*, which became Little Richard's unmistakable vocal signature; in Alex Bradford's pathos and understated irony woven into the country blues of Ray Charles; and in the fiery vocal pirouettes of Clara Ward that embellished the near-otherworldly artistry of Aretha Franklin at her soulful crest.

Three recent releases, *The Great 1955 Shrine Concert* and *The Great Gospel Men* and *The Great Gospel Women*—drawn largely from gospel's highest moment—confirm gospel's supreme ability to transport us to the heights of unalloyed joy and inspiration, even as its singers at times toiled in the valley of denial and disappointment.

The Shrine Concert, taped at the Shrine Auditorium in Los Angeles, and noteworthy for preserving live the electric emotions evoked onstage by the greatest gospel artists, captures—at their heights—gospel quartets like the Soul Stirrers (featuring Sam Cooke) and the Pilgrim Travelers, groups distinguished for tight harmonies and whirling melodies, delivered mostly in haunting a capella. On the Pilgrim Travelers's "All the Way" and "Straight Street," their sweet meditations on religious discipline and unfettered commitment to the Almighty are ushered in on a wave of "walking rhythms"—anchored by Jesse Whitaker's uncannily precise and harmonically creative baritone—for which they are justly famous.

Also present at the *Shrine Concert* were the Caravans, featuring the young James Cleveland and gospel diva Albertina Walker; the group's emotional reworking of Alex Bradford's brooding ballad "Since I Met Jesus" is an unexpected revelation. And Sam Cooke's soaring tenor—more gritty than on most of his later pop recordings—demonstrates that his most brilliant artistry occurred in the sacred space of the stage-as-sanctuary. The feral Dorothy Love Coates and the Original Gospel Harmonettes (contributing an explosive medley) and the incomparable Joe May round out the live recording.

Anthony Heilbut, America's foremost gospel-music authority, has performed an invaluable service by producing *The Great Gospel Men* and *The Great Gospel Women*, two albums that, in many cases, provide posthumous recognition for innovators of superior achievement. Most of the greats are included—from Mahalia Jackson, gospel's undeniable queen, with her bluesy rendition of "Just Keep Still" to Roberta Martin, who renders a soul-sundering serenade on "What a Friend (We Have in Jesus)."

On hand, too, is Clara Ward, Aretha Franklin's idol, who, on a late performance teased from her by Heilbut, unleashes volcanic eruptions of vocal intensity as she travels—carried along by her clear, sweet, nasal tones—"The Last Mile of the Way." And Marie Knight's "Walk with Me" caressed by her smoky contralto, transforms a sorrow song from slavery into a jaunty, jazzy, guitar-based plea for divine companionship.

The male stars receive their full due as well. The brilliant Brother Joe May, in a jewel of a live performance, illuminates both the raw passion and rhetorical complexity that join gospel singing and preaching on the classic "Move On Up a Little Higher." Robert Anderson, the equal of any blues singer, displays his uncanny timing (he often sings behind the beat) and a piercing, poignant baritone on the funereal organ chords of his classic "If Jesus Had to Pray." The spectacular J. Robert Bradley—considered by many the greatest male gospel soloist—shares his stunning gifts on "Pilgrim of Sorrow": He gently but firmly embraces the song's inherent sadness and explores the interior range of mood enveloped in its lyrics, displaying a consummate mastery of melismatic range with a Paul Robeson baritone wedded to a passionate gospel experience.

The late, great Marion Williams is amply represented on *The Great Gospel Women* by six tracks—a fitting recognition of both her protean talents and her splendid command of gospel genres, from bluesy ballads to field shouts. Williams's majestic voice truly can whisper or thunder with total conviction. The MacArthur "genius" grant that Williams was awarded a year before her death is not only a well-deserved acknowledgment of her individual talent but a long-overdue gesture of appreciation for the monumental contributions of all the artists represented here to the rich heritage of American music.

Mariah Carey
and "Authentic" Black Music

At its best, pop music presses an anxious ear to American society, amplifying our deepest desires and fears. At times, too, pop music almost unconsciously invites us to listen to ourselves in ways forbidden by cultural debates where complexity is sacrificed for certainty. In this vein, the reascent to the top of the charts of Mariah Carey's most recent album, *Music Box,* signals more than her musical dominance.

One source of Carey's significance—and undoubtedly the sharpest controversy around her—has nothing to do with the singer's gargantuan musical gifts. Instead it derives from the confusion and discomfort that her multiracial identity provokes in an American culture obsessed with race. Though she has made no secret that she is biracial (her mother is white, her father a black Venezuelan), Carey's candor evokes clashing responses from fans and critics. Some see her statement of mixed heritage as a refusal to bow to public pressure to choose whether she is black or white. But in light of the "one drop" rule—where a person is considered black by virtue of having one drop of black blood, a holdover from America's racist past—many conclude that the issue of racial identity, for Carey and other interracial people, is settled.

To make matters more complex, Carey's vocal style is firmly rooted in black culture. It features a soaring soprano and an alternately ethereal and growling melisma that pirouettes around gospel-tight harmonies. So if she's not clearly black yet sings in a black style, is she singing black music? And what difference does it make? Without even trying, Carey's music sparks reflections about how race continues to shape what we see and hear.

Partly what's at stake is the messy, sometimes arbitrary, politics of definition and categorization. What makes music "black music" and who can be said to legitimately perform it? Consider the fiery fusion of rock, soul and blues performed by Lenny Kravitz (like Carey, the child of an interracial marriage) and Terrence Trent D'Arby, or the socially conscious hard rock of the group Living Color. Is theirs black music? Though the answer is often negative, the roots of their music can be traced to black cultural influences, from Howlin' Wolf to Jimi Hendrix. The difficulty of fixing labels on what D'Arby, Kravitz, and Living Color do highlights the racial contradictions at the center of contemporary popular music.

Behind this painful, often protracted struggle to get at the "original article" is what can only be termed the anxiety of authenticity. Such quests are more than academic for black folk because of the history of appropriation and abuse of black musical styles by white performers and producers. While black artists like King Oliver and Chuck Berry initiated musical innovations from jazz to blues-based rock & roll, the public recognition and economic benefits due them evaporated, while derivative white artists like Benny Goodman and Elvis Presley reaped huge artistic and financial rewards.

Curiously enough, the debate over authenticity lies at the heart of hip-hop, though irreverence and transgression are staples of rap culture. But authenticity, even in a genre as closely identified with black culture as rap, does not strictly follow the rules of race. For instance, while the white rapper Vanilla Ice was greeted within hip-hop with derision because he came off as a white boy trying to sound black, white rap groups like Third Bass and House of Pain have been enthusiastically embraced because of their "legitimate" sounds and themes. Conversely, black rap artists like Hammer and the Fresh Prince have been widely viewed as sellouts because of their music's pop propensities.

An even thornier issue is the belief in black communities that some artists obscure their racial roots in a natural but lamentable response to a racist environment. As a result, they benefit from being black (given the extraordinary popularity of black music) but do not identify with the black people who support them before they discover a crossover market. In the extreme, this circumstance leads to the ideal of the pan-ethnic, omni-racial artist, an exotic fantasy whose energy derives from an implicit denial of the inherent value of simply being black. While Carey has been scrupulous at award ceremonies to thank her black fans, and to mention her black father in interviews, artists like Paula Abdul (a self-described "Syrian-Brazilian-Canadian-American" who first gained public notice as a "black" cheerleader/choreographer for the Los Angeles Lakers) have increasingly underplayed their black heritage.

Still, as the old saying goes, the finger pointed at artists implies several others pointed back at ourselves. American culture is painfully redefining itself through bitter debates about "identity politics," "multiculturalism," and "universalism." Music cannot be naively expected to triumph over social differences. Because of the schmaltz that often passes for conscience in pop, the dream of transcendence— whether of race, or for that matter, of sex and class—is often hindered by sappy appeals to brotherhood and oneness. What such impulses reflect is a desire to fix what has gone wrong in a culture intolerant of difference.

The anxiety of authenticity about what and who is really black in pop music is proportional to just how increasingly difficult it is to know the answer. As multiracial unions of sex and sound proliferate, the "one drop" rule may lose its power. And, as cultural theorists are now proud to announce, race is not merely a matter of biology but an artifice of cultural convention. Such a construction is often used to establish and reinforce the power of one group over another. This view does not mean that black music is solely the product of perception. Nor does it mean that black music's power must be diluted to a generic form. What it does suggest, however, is that the meaning of race, like the art it molds, is always changing.

In the end, what Carey's career may teach us is that paranoia about purity is the real enemy of black cultural expression, which at its best is characterized by the amalgamation of radically different elements. Creolization, syncretism, and hybridization are black culture's hallmarks. It is precisely in stitching together various fabrics of human and artistic experience that black musical artists have expressed their genius.

Old School Love

Chante Moore and Tony Terry

On their recent albums, Chante Moore and Tony Terry prove they are artists whose musical styles elegantly unite the classic and the contemporary in black pop. Moore's *A Love Supreme* and Terry's *Heart of a Man* explore the pleasures and thwarted ambitions of romance. Each artist evokes love's vibrant hues, its gritty and graceful textures, by subtly threading old-school passion through updated rhythms. The results are not only musically compelling, but they perhaps unintentionally illumine charged debates about the role of sex in black music.

As on Moore's 1992 debut album *Precious*, *A Love Supreme*, especially, measures the distance between explicit sexuality and the sophisticated romanticism that shapes her restless pursuit of perfect love. On "My Special Perfect One," wisps of flute float above a hypnotic, spiraling midtempo groove as her yearning soprano enumerates the qualities of an ideal mate. "I'm What You Need," an exquisite, breathy ballad, finds Moore exulting in her ability to please as feathery background harmonies slip through the song's percussive rhythms. And on the jazzy, hip-hop influenced "This Time," Moore passionately expresses gratitude for a second chance at getting love right.

At first blush Moore's romantic quest seems, well, uncomfortably traditional, the sort of knight-in-shining-armor idealism scored by feminist critics. For instance, on "Searchin'," which announces themes she later elaborates, Moore declares: "I believe there's someone for each of us/ So I've got to be at the right place/ At the right time, in my black pumps/ I check the mirror/ Although I see myself clearly/ It's you I can't find . . ."

But such concerns are allayed when one considers the misogyny of rap and the explicitness of much of today's rhythm-and-blues. Additionally, Moore imaginatively recasts the traditions she makes use of. She exploits the nostalgia for the lost art of inference, where carnal longings were not so much paraded as skillfully evoked. On "This Time," she sings: "And when you kiss me/ I'm gonna taste my lips/ I won't let go/ And when you hold my hand/ Baby, I won't say no."

She takes listeners forward by looking backward, giving new meaning to familiar songs. Moore weds the tag from Lionel Richie's 1979 anthem "Sail On" ("I've got to sail on honey/ Good times never felt so good") to the end of Deniece Williams's 1976 classic "Free" ("I want to be free/ And I've just got to be me"). The effect of this bit of revisionist magic is to sharpen the plea for independence only hinted at in Williams's lyrics. Moore's plaintive vocal twistings accentuate the rollicking "I Want to Thank You," but her gospel inspired coda underscores the religious passion the song draws from.

Perhaps the synergy of revived romanticism and propulsive contemporary rhythms is most dramatically realized on "Old School Lovin'." It is a sparkling, guitar-driven ode to intimacy built on the rituals of courtship and Moore's entrancing, overdubbed vocals: "The good times we had/ Somehow strayed away/ Let's put our love back/ Put lovin' back/ To how it used to be/ In the days of true romancing." The song's narrative energy derives from the sly, shifting metaphor at its center; "Old School Lovin'" speaks to the need to recapture the romantic legacy in relationships by both individuals and the society as a whole.

Terry's *Heart of a Man* covers similar terrain from a constantly self-reflective, and dare it be said, vulnerable masculine perspective. "When a Man Cries" lays claim to the sensitive side of the male ego. "There's nothing more tender," he proclaims in his poignant tenor, "Than the warm and loving tears/ Of a man." On the album's title single, Terry's pleasing falsetto layers his trademark yodel around the lyrics of a striding, horn-based ballad: "The heart of a man/ Is a big responsibility/ Seems so strong/ But it breaks so easily."

By these and other songs ("Can't Let Go" and "I'm Sorry"), Terry registers a rebuke to the woman-bashing found in so much hip-hop and rhythm-and-blues. His artistic persona is shaped by an aesthetic of male redemption that eschews female demonization and embraces the comforts and conflicts of commitment. This attitude lends authority to Terry's melismatic plea, issued over the jazzy but menacing rhythm of a song whose title says it all: "Don't Give Up a Good Man." It also bolsters the integrity of his assertion, made on the atmospheric ballad "Surrender," that "Girl I would never hurt you/ Cause hurting you is hurting me."

Like Moore, Tony Terry creatively engages the music that has inspired him. His version of Al Kooper's bluesy standard

"I Love You (More Than You'll Ever Know)" approaches the brilliance that made the song a minor classic for Donny Hathaway in 1973. Terry's effort throughout *Heart of a Man* more fully realizes the enormous potential only spottily revealed on his two previous albums.

On these two fine albums, Moore and Terry have certainly heartened those who believe that love is about more than butts, boobs, and knocking boots. Their art is ample evidence that today's artists have not only learned a lot, but they have valuable lessons to teach as well.

22

Crossing Over
Without Going Under

Luther, Anita, and Vanessa

The attempt to translate R&B achievement into pop success often involves a cruel irony: The most compelling features of black music are often compromised by artists' crossover ambitions. Fortunately, recent albums by Luther Vandross, Anita Baker, and Vanessa Williams, three standard-bearers of contemporary black pop, offer ample evidence that even as you venture from home base, you can keep the fire that defined your appeal.

Vandross's *Songs* especially illumines the encompassing visions and ironic quests that fuel and frustrate his career. Vandross has not easily crossed over to pop success because his craft is edifyingly specific, rooted in the particular nuances of the soul tradition in which he was nurtured. This album of remakes of well-known songs enlarges Vandross's delightful fetish for "Lutherization," innovatively shaping a classic song with his unique vocal signature. Vandross's vocal style weds the emotional range and stylistic sensibilities of female artists like Aretha Franklin to the hopeful romanticism outlined in the masculine fluidity of Sam Cooke.

On "Killing Me Softly" (Roberta Flack), "Reflections" (Diana Ross and the Supremes), "What the World Needs Now" (Dionne Warwick), "Since You've Been Gone" (Franklin), and "Evergreen" (Barbra Streisand), Vandross recasts lyrics originally interpreted by a pantheon of his female influences. He invests the songs with shades of meaning through ad-libs and melismas, sonically embellished by orchestral arrangements, gospel harmonies, and dense vocal layerings of his own voice. On "Killing Me Softly" he counterpoises the lyric "And then he looked right through me/ As if I wasn't there" with the aside "I was right there," further personalizing an already intimate revelation. And on "Since You've Been Gone," Vandross reinforces the plea for a departed lover's return with thickly textured vocal support accented by the singer's own witness.

Vandross's exploration of male, duo and group territory is equally intriguing. His gospel-tinged version of Stephen Stills's "Love the One You're With" ironically highlights the song's decidedly secular admonition to love by convenience more than principle. And his reprise of McFadden and Whitehead's inspirational "Ain't No Stoppin' Us Now" houses up the feel-good exhortation to overcome negativity. Although his treatment of Lionel Ritchie's "Hello" is breathtaking, faithful without being merely imitative, Vandross's earthy treatment of Rod Temperton's "Always and Forever" is too subdued; Vandross's bartione fails to match the ethereal rendition of the song by Johnnie Wilder. While there may be no stunning surprises on *Songs,* as with his past efforts at reworking a song's conceptual and musical limits

(think of his groundbreaking work on the Carpenters'
"Superstar" and Brenda Russell's "If Only for One Night"),
this collection, brilliant in many spots, proves that Vandross
is a master of musical reinterpretation.

Anita Baker interprets a few classics as well, though her
focus on *Rhythm of Love* is on enriching her artistic vocabulary while expanding her musical base. Baker's covers of "The
Look of Love," "You Belong to Me," and "My Funny Valen-
tine" are spare, powerful reworkings that testify to her considerable vocal strengths. When she wraps her voice around
an intelligent lyric, her smoky contralto has few equals. Her
artistry demands serious, repeated listening, and on the original songs presented on *Rhythm*, she delivers.

"Body and Soul" allows Baker to explore the contours of
total commitment, her rich, resonant tone riding tight gospel
harmonies. "Sometimes I Wonder Why" is characteristically
jazzy, a mellow meditation about the risks and limits of love
poignantly pleaded over Joe Sample's breezy piano chords.
Baker even exposes the once-taken-for-granted relationship
between R&B and country as stinging guitar chords accent
her searing profession of inexhaustible fidelity on "Plenty of
Room." But the real gem is "I Apologize," a haunting confession of culpability that goes right to the heart. Gordon Cham-
bers's layered background vocals, stretched between solace
and sincerity, are the perfect accompaniment to Baker's soaring, elegiac lament.

If Vandross and Baker have taken to reinventing classics,
Vanessa Williams has made a career of reinventing herself.
From deposed Miss America to multifaceted recording artist
and actress on film, television, and stage, Williams's many
successes have made her past troubles seem a necessary step
to achievement. Her first two albums imaginatively experi-
mented with dance grooves and brilliantly exploited her
inherent sensuality as a performer. (This last move was
shrewd, even courageous. For starters, Williams successfully
transformed and repackaged the sexuality that had drawn

derision from her critics when she posed nude. But with the fearless and critically celebrated embrace of a sexual self in her music, Williams has flouted arbitrary conventions that bring stigma to some artists while leaving others unscathed.)

Both of Williams's previous outings proved that her career was neither a fluke nor a novelty musical footnote. And now, just when she might have been expected to stick to familiar territory, Williams pulls up musical stakes on her third album, *The Sweetest Days,* boldly laying claim to a more acoustic esthethic that allows the natural beauty of her voice to shine.

The Sweetest Days is full of hidden pleasures and unexpected delights. Williams largely calls upon jazz to expand her musical vocabulary, but she also includes folk, pop, and soul. Even as Williams explores a jazz idiom, however, she artfully embraces its varied shades and emphases. For instance, on "It's the Way You Love Me," Williams pays tribute to her lover's charms with a mellow, bouncing groove that gently interweaves jazz, hip-hop, and soul. "Ooh, you know me so well/ You set my soul on fire/ And I know you got my rhythm/ 'Cause you take me higher," Williams sighs as a psychedelic guitar riff counterpoints a stuttered rap that underscores her theme. On "Ellamental" Williams more robustly and creatively fuses jazz and hip-hop. With bluesy, muted horns, layered vocal harmonies, jolts of sax, and light but skillful scatting, Williams crafts a delicious ode to Ella Fitzgerald's inimitable vocal style. Intriguingly, the tune also helps to clarify the musical kinship between jazz and rap, as a rapper declares on the song that "Listenin' to hop-hop/ Listenin' to be- bop/ And jazz/ Ella made you move that ass."

On her deft treatment of Sting's "Sister Moon," Williams invokes a subtle, cool spirit to sweetly swing the Brit's brooding number. And Williams's simple but elegant rephrasing of Patti Austin's "You Don't Have to Say You're Sorry," throws fresh light on a minor pop jazz classic. Austin's wise words sink right to the core of one of love's greatest ironies: the unspoken gift of forgiveness that sustains a mature relationship must simultaneously be taken for granted and explicitly sought out. "Explanations aren't required/ Even though you think you should/ You don't have to say you're sorry/ But I sure do wish you would," Williams tenderly intones.

That same tenderness surfaces more poignantly on the album's title song. "The Sweetest Days" aims consciously (perhaps too consciously) for the big ballad feel that fueled Williams's huge hit (with the same team of songwriters) "Save the Best for Last," from her second album. Williams is quite believable as a pop chanteuse. Her measured delivery gracefully plumbs the song's exhortation to savor life and love in the present.

Williams's stripped bare textures work effectively on two very different songs. On "Higher Ground," Williams explores the spiritual transformation invoked by authentic love in a slow winding folksy ballad built on the generous surge of acoustic guitar. And on the Latin tinged "Constantly" Williams probes her emotional ambivalence in coming to grips with the age old dilemma of falling for someone else's lover. Williams's vocal confidence is apparent; it is heard in her zesty rise to her upper register, and in the spare aural landscape against which she pitches her sweet serenade. The frets of a single guitar, overdubbed vocal harmonies, and grinding organ chords are her sole accompaniment.

But it's on the two Babyface tunes that Williams melds her varied artistic energies—Broadway actress and credible songstress—into a single dynamic. Drawing from both the sultry persona of Aurora (the sexy stage character she played in *Kiss of the Spider Woman*) and the sheer feistiness that has propelled her career forward from infamy, Williams finds musical complement in Babyface's occasionally acerbic meditation on love's vicious twists of fate.

On "Becha Never," a Brazilian feast spiced by Spanish guitars and airy vocal textures, Williams acidly asserts that revenge is the answer to infidelity. "I've found another/ And

he is my lover/ And I'm going out for the rest of your life/ I
becha never thought I'd go that far. . ." Williams taunts. "You
Can't Run" breezily sambas across a musical landscape composed of sparkling guitars, entrancing harmonies, and vibrant
percussion. In a compelling lower register Williams assures
her potential suitor that the lure of love is undeniable. "Ooh,
and you can try to run but there's nowhere to hide/ Love is
stronger than your lies," Williams insists. On *The Sweetest
Days* Vanessa Williams not only continues to brilliantly rede-
fine her musical identity, but she proves that past pain can be
the spur to substantive art.

Vandross, Baker, and Williams have helped define black
American music during the last decade. Their strengths and
weaknesses, successes and failures, index much more than
their individual art. Their work embodies the possibilities of
black pop, and it is to their credit that they continue to search
for the best expressions of their consistently engaging muses.

23

A Day in the Life
of Black Culture

A List

I. THE GOOD

1. Love Child. Before the release of his brilliant album *Heart, Mind & Soul,* you knew El Debarge had the goods: a sweet falsetto of remarkable clarity and grace. But too often he scaled to ethereal vocal heights with hit-or-miss material. His career cried out for a towering musical achievement to clarify the basis of his evasive appeal. By teaming with Babyface, the Midas-touch maestro of postmodern techno-soul, Debarge shapes an offering of love songs that pays imaginative homage to Marvin Gaye's hypnotic sensuality.

2. Detroit Soul City. The Motown Museum is chock full of memorabilia drawn from an era when black entrepreneurial ingenuity met transcendent talent. Together, these forces transformed music into history in a city whose cultural and economic motors were not yet stalled by postindustrial collapse, gentrification, car-jacking, white flights, and black tracks to the suburbs, and political corruption. Those small rooms and humble instruments, which produced such big results, remind me of just how much of a David-versus-Goliath story Motown really was. And seeing Berry, Michael, Diana, Marvin, Smokey, and Claudette, the Temps, the Four Tops, and the rest of the crew in black-and-white photos, recalls an era of faux innocence tragically shattered later in de-Africanized faces, infanticide, egomaniacal ambition, ripped-off royalties, and chemical slavery.

3. Fear of a Black Hat. I first saw this brash, witty film a few years ago when I spoke about violence in cinema at the Sundance Film Festival. I was disappointed to later learn that it was delayed, perhaps by the prospects of competing with the ever dumb, thoroughly crass, unrelentingly unsubtle *CB4*. *Fear* is on the money—both in its irreverent swipe at rap pieties (even if they do seem a bit dated by now); and in its absence of angst about mutating from the comedic gene of its obvious predecessor, "Spinal Tap."

4. Mama Said Knock You Out. The Salt-N-Pepa/R. Kelly tour was a smoldering, sumptuous blend of horniness and holiness. Smack dab in the middle of Kelly's sexual pyrotechnics—his stage show a throwback to the performer-cum-revival preacher on a mission to save souls—a picture of his mother descended. Kelly transmuted sheer erotic energy into spiritual pathos by remolding the Spinners classic "Sadie," a soaring elegy to black "mammas." That, plus Salt-N-Pepa's kick-ass feminism framed by in-heat gyrations fused to sanctified sexiness, make this not-to-miss show an orgasmic high of amens and oo-aahhs.

5. Girl's Got It Going On. With all the sound and fury over hip-hop these days, most people forget the central element in these debates: the music! With loads of intelligence and critical insight, Tricia Rose's book *Black Noise* offers a compelling analysis of the history and development of hip-hop as both musical and artistic expression. But she doesn't forget to pay attention to the politics of culture that shape its reception and interpretation. Necessary reading for pundits, professors, and politicians, but most of all, for those who love hip-hop's rhymes and reasons.

II. THE BAD

7. Andy Rooney on Kurt Cobain. Right, just what we need, another offering of transcendent wisdom from the man who has often turned curmudgeonly commentary into a (he)artless attack on the already hurt or dead. Rooney's vicious tirade against Cobain and a generation (black and white) already knocked unfairly for its absences—of spirit, industry, talent, and ambition—only reinforces the suspicions of too many young people that not only does a whole slice of this society not understand their plight, but it doesn't care either! Smells like mean spirit.

8. Rage against Rap, Part 1: U.S. Senate Hearings on Gangsta Rap. I participated in these. I can understand how fellow participants Dionne Warwick and C. Delores Tucker want to curb the attack on women in rap. Its lethal sexism and misogyny need to be resisted with all our might. But I can't figure out why or how a congressional committee is the best route to redemption, especially when that august body is a bastion of gender oppression. (It's why Senator Carol Moseley-Braun ran for the Senate in the first place; not because of Snoop Doggy Dogg, but because of Arlen Specter!) Plus, Ms. Warwick, I think far more harm is being done to America by celebrities snooping around the stars and selling answers to the perplexing questions of life—like will I ever get laid again, or will I win the lottery this week—rather than figuring out what political collapse has to do with cultural expression. Besides, didn't you see it coming?

9. Rage against Rap, Pt. 2: Playthell Benjamin on Thirtysomething Scholars. After a conference on race at my graduate school alma mater, Princeton, journalist Benjamin dissed me and Robin D. G. Kelley in his *New York Daily News* column (calling me a "snake-oil salesman" and accusing us of not knowing our history). The real deal seemed to be that Benjamin castigated us for the nerve we had to talk about young black folk without pathologizing them as the thugs and throwaways he apparently believes them to be. Let me break this one down for you: its old-school Negro skepticism at its ad hominem worst, worrying out loud whether young black folk are trustworthy agents to carry the genius of the race forward. More to the point, its the crotchety cry of jealousy aimed at brothas with flavor who have learning and burning, who refuse to demonize (or for that matter, romanticize) the very poor and working black people who gave us desire and reason to get Ph.D.'s in the first place.

10. "Kiss My Ring." Okay, give Jeru tha Damaja credit for at least wanting to make distinctions in his single, "Da Bichez." Of course, the problem is that it's still a man—relying upon the tried and true practice of surveillance and the male privilege of definition—who wants to determine for a woman what kind of female she should be. He also supplies the conditions she must satisfy in order to be counted either as a "sistuh," a "lady," a "woman," or a "bitch." Some sisters, ladies, and women might, ironically, prefer the rancid, ridiculous, but honest cant of Snoop's undifferentiated demonolgy: one man's "bitch" is another man's "bitch."

24

Hip-Hop and the Bad Rap

Hammer and Vanilla Ice

In hip-hop—where reputation is everything and the criteria for determining what's "real" changes in less time than it takes to fire a burst from an Uzi—the worst label to be stuck with is "sellout." Mind you, that tag is notoriously loose and inconsistently applied by artists and fans alike, and is often conveniently pinned on a rap rival or a despised star without rigorous justification. How do you rebound from such a tag, or, more important, how do you get your revenge if you're the artist who's been dissed?

Recent releases from two superstars who've been maligned for their pop propensities—Hammer's fourth album, *The Funky Headhunter,* and

Vanilla Ice's third, *Mind Blowin'*—contain intriguing if contradictory answers. The artists' strategies are markedly dissimilar; where Hammer blasts a sizzling rejoinder full of signifying and prideful parody into the ears of his accusers, Vanilla Ice crumbles under the pressure to perform. His attempt to harden his image simply doesn't work.

Throughout *Funky Headhunter* Hammer luxuriates in crunching bass lines, prancing grooves and irresistible hooks, even as his staccato delivery reflects a sorely needed improvement in his rapping skills. He and his gaggle of producers continue to refine his signature flourishes, building thick textures around sampled loops mined from funk's past. On the track "It's All Good", an infectious boast about his selling power, Hammer samples Brick's "Dusic." On "Oaktown," asserting that Hammer has ghetto roots, it's Prince's "Get It Up." And the carnival of carnality unleashed on "Pumps And A Bump" is driven by George Clinton's "Atomic Dog."

Hammer's foraging through funk history in his raps is a sly way of reestablishing his ghetto authenticity, borrowing from Dr. Dre's feral fusion of the two genres. While Dr. Dre's music harnesses Parliament-Funkadelic beats to gangster lyrics and life styles, Hammer jumps back a generation to reclaim a forgotten archetype: the ghetto hustler who lived more by his wits than by his guns. The narrator of the song "Something 'Bout the Goldie in Me," draws on the character in the '70's blaxploitation flick, *The Mack*.

As piercing guitar riffs slash through sheets of throbbing bass, Hammer reminds listeners that Goldie's supreme quest is to make money. Hence, Hammer justifies his pop success as the fulfillment of Goldie's genius: getting paid and staying alive.

In this light, Hammer paradoxically adopts a more hardedged persona by asserting the propriety of his pop past. As the Gap Band's "Shake" is buried in the snaking rhythms of the Teddy Riley-produced "Don't Stop," Hammer raps: "Now if a song that I drop hits the top 'cuz its pop/ then its like that and you don't stop." And Hammer brings acerbic humor to his repeated raps about not being a sellout, taking no prisoners as he playfully scorns his critics. "Some claimin' towns that they ain't even from/ Some claim they hard but never shot a gun," he raps.

Hammer is finally triumphant on *The Funky Headhunter* because his artistry is rooted in rap fundamentals: entertainment and emolument. His detractors can't have it both ways: either rap is a complex art form that makes equal use of reality and fiction or else the claims of studio gangsters and middle-class wannabees that Hammer's music lacks authenticity are bogus.

But if Hammer refuses to be cowed by his detractors, Vanilla Ice takes such criticism, and himself, too seriously on *Mind Blowin'*. It's not that he's any less defiant of naysayers than Hammer is; one need only listen to "Fame," where Ice details the downside of his success as a bleating guitar plays around the famous David Bowie–John Lennon melody: "Now in the public eye, ya know I'm havin' to admit it/ You become a target for a whole lot of critics."

It's that his rhetorical skills and his sense of humor failed him just when he needed them most. That failure makes much of *Mind Blowin'* almost unintentionally cartoonish, the "nyah-nyah-nyah" of a kid who sticks his tongue out at people who are shooting real bullets his way.

For example, "Roll 'Em Up," yet another ode to the magic of marijuana—"I need some herb and spices/so I can feel nices"—seems not only dated but downright hokey, despite its bumping bass and snatches of an almost extinct snare drum. "Hit 'Em Hard," featuring an eerie soundscape and a stop-time rap that accuses Hammer of "wack" songs (the artists share a mutual animosity), becomes an unintended self-parody, given Ice's rap-lite skills. And on "The Wrath" Ice claims to be back on the map, bringing scornful fire to his critics, while moaning piano riffs provide percussive reinforcement of the chanted self-quotation, "Ice, Ice, Baby", removing all fear of harm.

Of course, *Mind Blowin'* has good moments, but these are

largely the sampled songs Ice features: "Funky Rhymes" fronts a screeching guitar riff from George Duke's "Son of Reach For It (The Funky Dream)." "Blowin' My Mind," touting Ice's love skills, features an inventive reworking of Hall and Oates's "Sarah Smile." And "Minutes of Power," a boast about Ice's ability to rock a mike, features an entrancing melange of the Isley Brothers's "Take Me to the Next Phase" and James Brown's "Cold Sweat."

Like Hammer, Vanilla Ice is defensive about his past success; but unlike Hammer, Ice isn't skilled in the art of rhetorical response. Hammer's defensiveness translated into an imaginative, humorous gauntlet that assails hip-hop pieties. Vanilla Ice's self-serious attack yields a sonically interesting, lyrically feeble album that will do little to solve his problems—either of image or art.

Public Enemy

Rap's Prophets of Rage

Public Enemy is, hands down, the most influential and important group in the history of hip-hop. By roughly stitching together contrapuntal noise and prophetic rabble-rousing, the avant-garde group quickly became rap's conscience. The contrasting personalities of PE's duo—straight-man and heavy-duty lyricist Chuck D and trickster sidekick Flavor Flav—play off of one another to great effect. PE's work in toto has confronted, and at times embodied, most of the conflicts faced by young blacks over the last decade. Racist white media and sellout black bourgeoisie. Black-Jewish relations and the woes of interracial relationships. The narrowness of black radio and the betrayal of blacks by dope dealers. Through it all, PE has maintained its integrity and vision. This even as the group's themes—and popularity—have had to take a backseat to the mass appeal of gangsta rap in the 90's.

On *Yo! Bum Rush the Show,* PE's postmodern aesthetic and radical politics are only dimly foreshadowed. The nationalist agitprop that will later be proclaimed atop screeching guitars is gently paraded in *Rightstarter (Message to a Black Man).* On tracks like "Miuzi Weighs a Ton" there's a fair bit of boasting as well, mostly about Chuck's booming baritone and verbal skills, befitting the times in which PE rose to prominence. Still, what political gestures there were, often symbolic ones like Chuck D's willingness to declare himself a public enemy and the music's punkish commitment to deliberately irritating feedback scrapes and noise for its own sake, made Public Enemy stand out in the hip-hop world.

From its first words—a British voice introducing the group as if to indicate the essential foreignness of what's to come—to its final beats, the revolutionary *It Takes a Nation of Millions to Hold Us Back* lunges far beyond anything in rap's past to help secure its future. Just one year after Run DMC's landmark album *Raising Hell* cemented rap's commercial appeal, *It Takes a Nation* gave the genre ideological vitality. Bombastic beats and clashing polyrhythms ferociously leap off the soundscape, a tribute to production crew The Bomb Squad's orchestrated cacophony. On "Bring the Noise," Chuck's startling, irrepressible flow, punctuated by snippets of Flav's off kilter commentary, strikes close to home. "Radio stations I question their blackness/ They call themselves black, but we'll see if they play this/ Turn it up! Bring the noise!" And on "Don't Believe the Hype," Chuck and Flavor rap over a repetitive shrieking noise, intermittent scratching, a ghoulish glissando, and a breezy bass. Chuck proclaims himself the "follower of Farrakhan/ Don't tell me that you understand/ Until you hear the man." As the song boomed out of car stereos across the United States in the summer of 1988, it was clear the group's leader had truly accomplished his stated goal: "Teach the bourgeoise, and rock the boulevard."

PE consolidated its reputation as hip-hop's preeminent voice of prophetic rage on *Fear of a Black Planet.* Cuts like "Brothers Gonna Work It Out" ("Brothers. . . get mad, revolt, revise, realize"), "Anti-Nigger Machine" ("Never had to be bad/ My Mama raised me mad"), and "Revolutionary Generation" ("It's just a matter of race/ Cause a black male's in their face") continue to craft PE's resistance to white racism—and embrace of white noise. The group faced charges of anti-Semitism over the lyrics to "Welcome to the Terrordome": "Crucifixion ain't no fiction/ So called chosen frozen/ Apology made to who ever pleases/ Still they got me like Jesus." While such sentiments marred their revolutionary agenda, PE set its healthier strivings to a vibrant musical mix-and-match that prominently features funk rhythms. "Fight the Power," theme song for Spike Lee's cinematic PE equivalent, *Do the Right Thing,* illustrates this nicely, propelled by the irresistible energy of a coiling James Brown shuffle and samples of Aaron Hall's funky tenor.

PE was more thematically consistent than ever on *Apocalypse 91 . . . The Enemy Strikes Black,* with jeremiads against drinking ("1 Million Bottle Bags"), black radio ("How to Kill a Radio Consultant"), and internalized self-hatred ("I Don't Wanna Be Called Yo Niga"). Or was the group just repeating itself? By 1991 the rules and rulers of hip-hop were changing, with biting black nationalist commentary and an Afrocentric worldview giving way to sexual hedonism and the glamorization of violence. Public Enemy failed to react. Still, there is punch to "Can't Truss It," a tribute to black survival in a wicked white world. And the brilliant wordplay of "By the Time I Get to Arizona," is used to attack the refusal of Arizona's then-Governor Meacham to recognize Martin Luther King Jr.'s birthday as a state holiday.

Greatest Misses is pretty evenly split between new stuff

and remixes. Though this album was almost universally pooh-poohed, it contains some interesting moments. "Tie Goes to the Runner" works off an infectious jazz groove. On "Gett Off My Back," Flavor Flav showcases his trademark unorthodox rhyme skills and comedic talent. Public Enemy's rebirth in the Age of Gangsta is heralded on *Muse Sick 'N Hour Mess Age*. Because it is the group's first work in a few years, and records their responses to the near ubiquitous gangsta rap, I'll examine the album in detail.

II

What a difference a Dre makes. Especially in hip-hop where that wizardly stalwart of gangsta funk and his sonic siblings enjoy early '90s rap as their oysters. Fighting material misery with a battery of blunts and forty ounces, guns and gangs, "bitches" and hood-rats, gangsta rappers rudely depart from the aesthetics and ideology of their more politically conscious kin. The angry distance between the hoisted banner of carnal excess and its excoriation by animated if embattled prophets registers full-throttle on Public Enemy's fifth album, *Muse Sick 'N Hour Mess Age*. Public Enemy's rebirth is heralded by a new sound created from chanted choruses, sung refrains, and live (not sampled) trap drums. If this doesn't quite make them groove-meisters, it does mellow the musical blasts of former producers The Bomb Squad, who in the past stitched together a postmodern pastiche of rump-rattling noise.

Though Public Enemy's album may be easier to listen to, their message of racial apocalypse leaps from the record with undiminished urgency. Except now, they believe that too many of their own number are sleeping with the enemy. On several tracks, especially "Give It Up," and "So Whatcha Gone Do Now?" Chuck D leverages his legendary lyrical skills against what he sees as the moral bankruptcy and racial betrayal of gangsta rap. Rather than boozin' and blowin' joints, he'd rather "diss presidents, dead or alive," an assertion issued over a mellow electric piano riff weaved throughout "Give It Up." And over the mellifluous chaos of the Coltranesque "So Whatcha Gone Do Now," an unsettling colloquy of voices float in and out of hearing to complement and complete Chuck's lingering lament that "Everybody talkin' that drive-by shit/Talkin' that gangsta shit."

On "Whole Lotta Love Goin' On in the Middle of Hell," Chuck follows an ominous bass line through a minimal aural field as he insists that the real beneficiaries of gangsta rap are the denizens of the racist New World Order. And on "Bedlam" Chuck implores gangsta rappers to see the link between a socially sanctioned desire for self-destruction, failed ecological responsibility and the mindless escape encouraged by technological gadgetry as he admonishes against "indo/ Throw it out the window/ Along with the super-nintendo," warning that "The world can't take it no more, no more/ The earth gets treated like a whore." The track is a sly merger of sentiments expressed in separate cuts on Marvin Gaye's groundbreaking album *What's Goin' On,* as PE join an environmental ethic to a social conscientiousness with hip-hop flavor.

Undoubtedly some will view Public Enemy's harsh criticism of gangsta rap as a grudging failure to acknowledge just how much and how quickly times have changed. Others will think that it's PE's graceless refusal to bow out and admit that their day has passed, that other rappers must now follow their appointed paths. Still others may suspect that Public Enemy has no sense of humor or complexity, that for all their undeniable wisdom about black youth and their varieties of rage, that PE just don't get it. After all, signifying and play, distortion and hyperbole, lift studio gangsterism to an art. Worse still, others may believe that PE is getting conservative in its old age. Its ironic politics of revolt only reinforce the status quo, some may contend, placing PE in league with reactionary elements of the black bourgeoisie and white conservatives which the group has lambasted in the past.

Admittedly, parts of these charges may in some abstract sense ring true; but their overall impact is lessened by paying attention to the album's overarching ambition and to hip-hop's perennial quest for change, its virtually unalterable mandate for renewal. Chuck D's diss of gangsta rappers departs from their bashing at the hands of unthinking naysayers because his criticism is rooted in a radical political project of racial redemption.

On "Ain't Nuttin' Butter Song," punk-funk rhythms are accented by screeching Hendrix-like guitar riffs on "The Star Spangled Banner" as Chuck D and Flavor Flav rap about how that bit of Americana is a jaundiced justification for racial tyranny and every bit as destructive as any violence imagined by gangsta rappers. On "Race Against Time," a bass groove crunches as scratches prick Public Enemy's evocations of prophetic common sense in understanding how plagues—from the infamous Tuskegee experiment to AIDS—have harmed and been identified with black people throughout the diaspora. And jolts of saxophone shower PE's indictment of the vulture-like ways of Eurocentrism on "Thin Line Between Law & Rape," its materialism and criminality infecting everything it seeks to conquer. In the political frame of PE's revisionist historicizing, their attacks on gangsta rap are consistent with criticisms of white oppression and capitalist exploitation.

Then, too, in the rapidly cyclical nature of hip-hop culture, what is needed just now is a dose of radical self-criticism that steals the thunder from would-be censors and uninsightful critics. (What's really funny is how PE diss critics on "I Stand Accused," when in truth what they're serving up squares with good old social criticism at its livid best.) In classic manner, PE's insistent, uncompromising demands, their cantankerous assertions of truth, their whole-boiled quest for justice fit the portfolio of all prophets possessed of a vision that just won't go away. Their discomfort rises from the irritating gift of sight laid on them, a burden that only increases when the times are inhospitable to their message. Like the Old Testament prophet Jeremiah, if they don't tell their story, its like a fire shut up in *their* bones. But when they do speak, agree with them or not, the flame burns us all.

Ice Cube

Gangsta Rap's Visionary

Ice Cube is to hip-hop what Prince is to R&B and rock: a prolific innovator of stunning talent with one big idea and lots of smaller ones. Ice Cube's obsession lies in lyrically unearthing the horrors and subversive pleasures of the South Central ghettoes he helped cloak in rap mythology. As part of the group NWA—the attending physicians at gansta' rap's 1988 birth on their pioneering *Straight Outta Compton*—Ice Cube's oral artistry was buttressed by Dr. Dre's ferocious G-funk grooves. Even when Ice Cube misses the mark, the furious intelligence and rhetorical skill of his gangstAfronationalist aesthetic manages to provoke and inspire. Breaking with NWA at an extraordinarily young age for one who had given

gangsta rap its most believable character, Cube has gone on
to imprint his blunt anger over several crucial solo albums.
He has consistently explored the self-contained doomy presence that also gained him critical acclaim apart from music in
the film *Boyz In The Hood,* and most recently, *Friday.*

On *Amerikkka's Most Wanted,* Ice Cube seizes the spotlight. On "The Nigga Ya Love to Hate," he is "kickin' shit
called street knowledge," demanding to know "Why more
niggas in the pen than in college?" The track is bolstered by
a swaggering bass line, the key to an aural assault that,
throughout the album, is presided over by Public Enemy's
Bomb Squad. Cube relentlessly exposes the terror of police
brutality ("Amerikkka's Most Wanted"), yet mirrors such
treatment in his own attacks on the fairer sex ("Once Upon a
Time in the Projects" and "I'm Only Out for One Thing").
Cube also gets delirious joy in rattling the pieties of the black
bourgeoisie; "Turn Off the Radio," a propulsive synthesis of
horns and psychedelic guitars acerbically blasts the R&B
lovers who in the late '80s crowded rap off the airwaves.

To view Cube's EP *Kill at Will* as filler is to miss a couple
of premises of hip-hop it helped to brilliantly illustrate: to
clarify a rap's intellectual content, remix it's musical elements;
and to test the marketplace, foreshadow what's to come. On
Kill at Will, Cube offers a foretaste of his more focused reflections on black male misery. He sets a mellow, jazzy tone that
will more frequently show up on his later albums. Commerce
and craft have rarely been more headily joined. The remixed
"Endangered Species" rides a wave of stinging guitars and
bombastic bass, as Cube and guest Chuck D engage in home-
boy ethnography, angrily asserting that black males are the
victims of social genocide. "Jackin' for Beats" is a shameless
apologia for musical thievery; Cube's stentorian style threads
the pastiche of rhythms he gleefully jacks from other rappers.
But the centerpiece is "Dead Homiez," an elegy to his fallen
friends, conversationally eloquent as bruising sax riffs echo
over the musical landscape.

Cube's sophisticated gangsta worldview made him a fellow

traveler to more politically explicit rappers like Public Enemy. This becomes obvious on *Death Certificate*. The album cover symbolizes a changed outlook, as his scowling countenance can be glimpsed shorn of his trademark jheri curls, a nod, no doubt, to his flirtation with Nation of Islam beliefs. *Death Certificate's* keenly contradictory morality play is fueled both by outrage at both black disdain for ghetto youth and by white racism. Cube rails against government corruption of black life ("I Wanna Kill Sam") and gang disputes ("Color Blind"), and presciently explores tensions between blacks and Koreans ("Black Korea").

What makes this album so brilliant and disturbing is its clarity. The volatile mix of acerbic social criticism and unvarnished bigotry (women represented as thwarting bitches or outlets for male lust; epithets like sissy, punk, and fag) provides a peek into the complex and problematic workings of an urban male on the margins of both black culture and white America. *Certificate* is split into "The Death Side" and "The Life Side," but one can scarcely trace any shifts in outlooks between them.

On his next album, *The Predator*, Cube is recycling the musical energies of his own brand of gangsta funk, if sometimes with expanded expertise. On "Wicked," for instance, a ragamuffin rapper's verbal accompaniment becomes a natural ally to Cube's staccato boasts. "It Was a Good Day," is an infectious, ironic allegory of a day in the life of a young black male, a Solzhenitsyn tale of terror in reverse. The rap's force derives from its celebration of what didn't happen—death, violence, police brutality—and from its irresistible anchor in the Isley Brothers's "Footsteps in the Dark."

By the time Cube's "Lethal Injection" was released, former mate Dr. Dre had not only produced the landmark album *The Chronic*, but he was presiding over the Hegelian eruption of gangsta Funk's Great Man of Rap, Snoop Doggy Dogg. "Really doe"'s eery, prickly aural soundscape reflects this influence. But what Cube gave up to Dre and company in sheer velocity of groove, he maintained in his unparalleled ability to fuse rage and political savvy. "Ghetto Bird," an apt metaphor for an L.A. police helicopter, captures the governmental surveillance of black urban life. Not that Cube couldn't still groove; "Ghetto Bird," dances along a menacing, rhythmic shuffle. And "You Know How We Do It," brilliantly integrates a sample from Evelyn Champagne King's "The Show Is Over," measuring a soulful, midtempo pace that also pays tribute to Kool and the Gang's "Summertime."

Bootlegs & B-Sides showcases Cube's ability to recast his earlier musical ideas, giving them longer life and more flavor. His wily remix of "Check Yo Self" jumps back to primal hip-hop territory in its straight-up appropriation of Grandmaster Flash and the Furious Five's "The Message." "It Was a Good Day" floats from one seventies soul icon to another, fleeing the Isley Brothers's coop to nestle in the serpentine rhythms of the Staples Singers's "Let's Do It Again." And the remixed "When I Get to Heaven" gives an even darker tone to Cube's stark indictment of Christianity (backed by the Nation of Islam) for its sins against black folk. "They won't call me a nigga," Cube confidently insists, "When I get to Heaven." On this, as with all of his artistry, Cube refuses to be anything but bold, brash, and provocative.

27

Gangsta Rap
and American Culture

The recent attacks on the entertainment industry, especially gangsta rap, by Senator Bob Dole, former Education Secretary William Bennett, and political activist C. Delores Tucker, reveal the fury that popular culture can evoke in a wide range of commentators. As a thirty-five-year-old father of a sixteen year-old son and as a professor and ordained Baptist minister who grew up in Detroit's treacherous inner city, I too am disturbed by many elements of gangsta rap. But I'm equally anguished by the way many critics have used its artists as scapegoats. How can we avoid the pitfall of unfairly attacking black youth for problems that bewitched our culture long before they gained prominence? First, we should understand what forces drove the emergence of rap. Second, we should place the debate about gangsta rap in the context of a much older debate about "negative" and "positive" black images. Finally, we should acknowledge that gangsta rap crudely exposes harmful beliefs and practices that are often maintained with deceptive civility in much of mainstream society, including many black communities.

If the fifteen-year evolution of hip-hop teaches us anything, it's that history is made in unexpected ways by unexpected people with unexpected results. Rap is now safe from the perils of quick extinction predicted at its humble start. But its birth in the bitter belly of the '70s proved to be a Rosetta stone of black popular culture. Afros, "blunts," funk music, and carnal eruptions define a "back-in-the-day" hip-hop aesthetic. In reality, the severe '70s busted the economic boom of the '60s. The fallout was felt in restructured automobile industries and collapsed steel mills. It was extended in exported employment to foreign markets. Closer to home, there was the depletion of social services to reverse the material ruin of black life. Later, public spaces for black recreation were gutted by Reaganomics or violently transformed by lethal drug economies.

Hip-hop was born in these bleak conditions. Hip-hoppers joined pleasure and rage while turning the details of their difficult lives into craft and capital. This is the world hip-hop would come to "represent": privileged persons speaking for less visible or vocal peers. At their best, rappers shape the tortuous twists of urban fate into lyrical elegies. They represent lives swallowed by too little love or opportunity. They represent themselves and their peers with aggrandizing anthems that boast of their ingenuity and luck in surviving. The art of "representin'" that is much ballyhooed in hip-hop is the witness of those left to tell the afflicted's story.

As rap expands its vision and influence, its unfavorable origins and its relentless quest to represent black youth are both a consolation and challenge to hip-hoppers. They remind rappers that history is not merely the stuff of imperial dreams from above. It isn't just the sanitizing myths of those with political power. Representing history is within reach of those who seize the opportunity to speak for themselves, to represent their own interests at all costs. Even rap's largest controversies are about representation. Hip-hop's attitudes toward women and gays continually jolt in the unvarnished malevolence they reveal. The sharp responses to rap's misogyny and homophobia signify its central role in battles over the cultural representation of other beleaguered groups. This is particularly true of gangsta rap.

While gangsta rap takes the heat for a range of social maladies from urban violence to sexual misconduct, the roots of our racial misery remain buried beneath moralizing discourse that is confused and sometimes dishonest. There's no doubt that gangsta rap is often sexist and that it reflects a vicious misogyny that has seized our nation with frightening intensity. It is doubly wounding for black women who are already beset by attacks from outside their communities to feel the thrust of musical daggers to their dignity from within. How painful it is for black women, many of whom have fought valiantly for black pride, to hear the dissonant chord of disdain carried in the angry epithet "bitch."

The link between the vulgar rhetorical traditions expressed in gangsta rap and the economic exploitation that dominates the marketplace is real. The circulation of brutal images of black men as sexual outlaws and black females as "'ho's" in many gangsta rap narratives mirrors ancient stereotypes of black sexual identity. Male and female bodies are turned into commodities. Black sexual desire is stripped of redemptive uses in relationships of great affection or love.

gangsta rappers, however, don't merely respond to the values and visions of the marketplace; they help shape them as well. The ethic of consumption that pervades our culture certainly supports the rapacious materialism shot through the narratives of gangsta rap. Such an ethic, however, does not exhaust the literal or metaphoric purposes of material wealth in gangsta culture. The imagined and real uses of money to help one's friends, family, and neighborhood occupies a prominent spot in gangsta rap lyrics and lifestyles.

Equally troubling is the glamorization of violence and the romanticization of the culture of guns that pervades gangsta rap. The recent legal troubles of Tupac Shakur, Dr. Dre, Snoop Doggy Dogg, and other gangsta rappers chastens any defense of the genre based on simplistic claims that these artists are merely performing roles that are divorced from real life. Too often for gangsta rappers, life does indeed imitate and inform art.

But gangsta rappers aren't *simply* caving in to the pressure of racial stereotyping and its economic rewards in a music industry hungry to exploit their artistic imaginations. According to this view, gangsta rappers are easily manipulated pawns in a chess game of material dominance where their consciences are sold to the highest bidder. Or else gangsta rappers are viewed as the black face of white desire to distort the beauty of black life. Some critics even suggest that white record executives discourage the production of "positive rap" and reinforce the desire for lewd expressions packaged as cultural and racial authenticity.

But such views are flawed. The street between black artists and record companies runs both ways. Even though black artists are often ripe for the picking—and thus susceptible to exploitation by white and black record labels—many of them are quite sophisticated about the politics of cultural representation. Many gangsta rappers helped to create the genre's artistic rules. Further, they have figured out how to financially exploit sincere and sensational interest in "ghetto life." gangsta rap is no less legitimate because many "gangstas" turn out to be middle-class blacks faking home boy roots. This fact simply focuses attention on the genre's essential constructedness, its literal artifice. Much of gangsta rap makes voyeuristic whites and naive blacks think they're getting a slice of authentic ghetto life when in reality they're being served colorful exaggerations. That doesn't mean, however, that the best of gangsta rappers don't provide compelling portraits of real social and economic suffering.

Critics of gangsta rap often ignore how hip-hop has been developed without the assistance of a majority of black communities. Even "positive" or "nation-conscious" rap was initially spurned by those now calling for its revival in the face of gangsta rap's ascendancy. Long before white record executives sought to exploit transgressive sexual behavior among blacks, many of us failed to lend support to politically motivated rap. For instance, when political rap group Public Enemy was at its artistic and popular height, most of the critics of gangsta rap didn't insist on the group's prominence in black cultural politics. Instead, Public Enemy and other conscientious rappers were often viewed as controversial figures whose inflammatory racial rhetoric was cause for caution or alarm. In this light, the hue and cry directed against gangsta rap by the new defenders of "legitimate" hip-hop rings false.

Also, many critics of gangsta rap seek to curtail its artistic freedom to transgress boundaries defined by racial or sexual taboo. That's because the burden of representation falls heavily on what may be termed the race artist in a far different manner than the one I've described above. The race artist stands in for black communities. She represents millions of blacks by substituting or sacrificing her desires and visions for the perceived desires and visions of the masses. Even when the race artist manages to maintain relative independence of vision, his or her work is overlaid with, and interpreted within, the social and political aspirations of blacks as a whole. Why? Because of the appalling lack of redeeming or nonstereotypical representations of black life that are permitted expression in our culture.

This situation makes it difficult for blacks to affirm the value of nontraditional or transgressive artistic expressions.

Instead of viewing such cultural products through critical eyes—seeing the good and the bad, the productive and destructive aspects of such art—many blacks tend to simply dismiss such work with hypercritical disdain. A suffocating standard of "legitimate" art is thus produced by the limited public availability of complex black art. Either art is seen as redemptive because it uplifts black culture and shatters stereotypical thinking about blacks, or it is seen as bad because it reinforces negative perceptions of black culture.

That is too narrow a measure for the brilliance and variety of black art and cultural imagination. Black folk should surely pay attention to how black art is perceived in our culture. We must be mindful of the social conditions that shape perceptions of our cultural expressions and that stimulate the flourishing of one kind of art versus another. (After all, die-hard hip-hop fans have long criticized how gangsta rap is eagerly embraced by white record companies while "roots" hip-hop is grossly underfinanced.)

But black culture is too broad and intricate—its artistic manifestations too unpredictable and challenging—for us to be *obsessed* with how white folk view our culture through the lens of our art. And black life is too differentiated by class, sexual identity, gender, region, and nationality to fixate on "negative" or "positive" representations of black culture. Black culture is good and bad, uplifting and depressing, edifying and stifling. All of these features should be represented in our art, should find resonant voicing in the diverse tongues of black cultural expressions.

gangsta rappers are not the first to face the grueling double standards imposed on black artists. Throughout African-American history, creative personalities have sought to escape or enliven the role of race artist with varying degrees of success. The sharp machismo with which many gangsta rappers reject this office grates on the nerves of many traditionalists. Many critics argue that since gangsta rap is often the only means by which many white Americans come into contact with black life, its pornographic representations and brutal stereotypes of black culture are especially harmful. The understandable but lamentable response of many critics is to condemn gangsta rap out of hand. They aim to suppress gangsta rap's troubling expressions rather than critically engage its artists and the provocative issues they address. Or the critics of gangsta rap use it for narrow political ends that fail to enlighten or better our common moral lives.

Tossing a moralizing *j'accuse* at the entertainment industry may have boosted Bob Dole's standing in the polls over the short term. It did little, however, to clarify or correct the problems to which he has drawn dramatic attention. I'm in favor of changing the moral climate of our nation. I just don't believe that attacking movies, music, and their makers is very helpful. Besides, right-wing talk radio hosts wreak more havoc than a slew of violent films. They're the ones terrorist Timothy McVeigh was inspired by as he planned to bomb the Federal Building in Oklahoma City.

A far more crucial task lies in getting at what's wrong with our culture and what it needs to get right. Nailing the obvious is easy. That's why Dole, along with William Bennett and C. Delores Tucker, goes after popular culture, especially gangsta rap. And the recent attempts of figures like Tucker and Dionne Warwick, as well as national and local lawmakers, to censor gangsta rap or to outlaw its sale to minors are surely misguided. When I testified before the U.S. Senate's Subcommittee on Juvenile Justice, as well as the Pennsylvania House of Representatives, I tried to make this point while acknowledging the need to responsibly confront gangsta rap's problems. Censorship of gangsta rap cannot begin to solve the problems of poor black youth. Nor will it effectively curtail their consumption of music that is already circulated through dubbed tapes and without the benefit of significant airplay.

A crucial distinction needs to be made between censorship of gangsta rap and edifying expressions of civic responsibility and community conscientiousness. The former seeks to prevent the sale of vulgar music that offends mainstream moral sensibilities by suppressing the First Amendment. The latter, however, is a more difficult but rewarding task. It seeks to oppose the expression of misogynistic and sexist sentiments in hip-hop culture through protest and pamphleteering, through community activism, and through boycotts and consciousness raising.

What Dole, Bennett, and Tucker shrink from helping us understand—and what all effective public moralists must address—is why this issue now? Dole's answer is that the loss of family values is caused by the moral corruption of popular culture, and therefore we should hold rap artists, Hollywood moguls, and record executives responsible for our moral chaos. It's hard to argue with Dole on the surface, but a gentle scratch reveals that both his analysis and answer are flawed.

Too often, "family values" is a code for a narrow view of how families work, who gets to count as a legitimate domestic unit, and consequently, what values are crucial to their livelihood. Research has shown that nostalgia for the family of the past, when father knew best, ignores the widespread problems of those times, including child abuse and misogyny. Romantic portrayals of the family on television and the big screen, anchored by the myth of the Benevolent Patriarch, hindered our culture from coming to grips with its ugly domestic problems.

To be sure, there have been severe assaults on American families and their values, but they have not come mainly from Hollywood, but from Washington with the dismantling of the Great Society. Cruel cuts in social programs for the neediest, an upward redistribution of wealth to the rich, and an unprincipled conservative political campaign to demonize poor black mothers and their children have left latter-day D. W. Griffiths in the dust. Many of gangsta rap's most vocal black critics (such as Tucker) fail to see how the alliances they forge with conservative white politicians such as Bennett and

Dole are plagued with problems. Bennett and Dole have put up roadblocks to many legislative and political measures that would enhance the fortunes of the black poor they now claim in part to speak for. Their outcry resounds as crocodile tears from the corridors of power paved by bad faith.

Moreover, many of the same conservative politicians who support the attack on gangsta rap also attack black women (from Lani Guinier to welfare mothers), affirmative action, and the redrawing of voting districts to achieve parity for black voters. The war on gangsta rap diverts attention away from the more substantive threat posed to women and blacks by many conservative politicians. gangsta rap's critics are keenly aware of the harmful effects that genre's misogyny can have on black teens. Irionically, such critics appear oblivious to how their rhetoric of absolute opposition to gangsta rap has been used to justify political attacks on poor black teens.

That doesn't mean that gratuitous violence and virulent misogyny should not be opposed. They must be identified and destroyed. I am wholly sympathetic, for instance, to sharp criticism of gangsta rap's ruinous sexism and homophobia, though neither Dole, Bennett, nor Tucker have made much of the latter plague. "Fags" and "dykes" are prominent in the genre's vocabulary of rage. Critics' failure to make this an issue only reinforces the inferior, invisible status of gay men and lesbians in mainstream and black cultural institutions. Homophobia is a vicious emotion and practice that links mainstream middle-class and black institutions to the vulgar expressions of gangsta rap. There seems to be an implicit agreement between gangsta rappers and political elites that gays, lesbians, and bisexuals basically deserve what they get.

But before we discard the genre, we should understand that gangsta rap often reaches higher than its ugliest, lowest common denominator. Misogyny, violence, materialism, and sexual transgression are not its exclusive domain. At its best, this music draws attention to complex dimensions of ghetto life ignored by many Americans. Of all the genres of hip-hop—from socially conscious rap to black nationalist expressions, from pop to hardcore—gangsta rap has most aggressively narrated the pains and possibilities, the fantasies and fears, of poor black urban youth. gangsta rap is situated in the violent climes of postindustrial Los Angeles and its bordering cities. It draws its metaphoric capital in part from the mix of myth and murder that gave the Western frontier a dangerous appeal a century ago.

gangsta rap is largely an indictment of mainstream and bourgeois black institutions by young people who do not find conventional methods of addressing personal and social calamity useful. The leaders of those institutions often castigate the excessive and romanticized violence of this music without trying to understand what precipitated its rise in the first place. In so doing, they drive a greater wedge between themselves and the youth they so desperately want to help.

If Americans really want to strike at the heart of sexism and misogyny in our communities, shouldn't we take a closer look at one crucial source of these blights: religious institutions, including the synagogue, the temple, and the church? For instance, the central institution of black culture, the black church, which has given hope and inspiration to millions of blacks, has also given us an embarrassing legacy of sexism and misogyny. Despite the great good it has achieved through a heroic tradition of emancipatory leadership, the black church continues to practice and justify *ecclesiastical apartheid*. More than 70 percent of black church members are female, yet they are generally excluded from the church's central station of power, the pulpit. And rarely are the few ordained female ministers elected pastors.

Yet black leaders, many of them ministers, excoriate rappers for their verbal sexual misconduct. It is difficult to listen to civil rights veterans deplore the hostile depiction of women in gangsta rap without mentioning the vicious sexism of the movements for racial liberation of the 1960s. And of course the problem persists in many civil rights organizations today.

Attacking figures like Snoop Doggy Dogg or Tupac Shakur—or the companies that record or distribute them—is an easy out. It allows scapegoating without sophisticated moral analysis and action. While these young black males become whipping boys for sexism and misogyny, the places in our culture where these ancient traditions are nurtured and rationalized—including religious and educational institutions and the nuclear family—remain immune to forceful and just criticism.

Corporate capitalism, mindless materialism, and pop culture have surely helped unravel the moral fabric of our society. But the moral condition of our nation is equally affected by political policies that harm the vulnerable and poor. It would behoove Senator Dole to examine the glass house of politics he abides in before he decides to throw stones again. If he really wants to do something about violence, he should change his mind about the ban on assault weapons he seeks to repeal. That may not be as sexy or self-serving as attacking pop culture, but it might help save lives.

gangsta rap's greatest "sin" may be that it tells the truth about practices and beliefs that rappers hold in common with the mainstream and with black elites. This music has embarrassed mainstream society and black bourgeois culture. It has forced us to confront the demands of racial representation that plague and provoke black artists. It has also exposed our polite sexism and our disregard for gay men and lesbians. We should not continue to blame gangsta rap for ills that existed long before hip-hop uttered its first syllable. Indeed, gangsta rap's in-your-face style may do more to force our nation to confront crucial social problems than countless sermons or political speeches.

Benediction

Letter to My Wife Marcia

Dear Marcia:

I've been thinking about our relationship, my
relationships in the past, relationships in general,
a lot lately. That's because as I travel the country
lecturing to dramatically different audiences—
college students and political groups, church
gatherings and corporate settings—the issue of
how black men treat black women, and vice versa,
invariably crops up. I've even done my fair share
of lectures, conflict resolutions and debates
devoted especially to "black male-female relations," a cottage-industry of black angst mined by
experts and exploiters alike. What bothers me
most, though, is the obvious psychic pain of men

and women, and teens too, who struggle under the burden of thwarted intimacy. The same folk frequently fail to realize how sex and social power are often fatally yoked.

In my experiences at forums addressing "black male-female relations," the display of gender trouble often follows a predictable pattern. At first, the hurts of black men and women are thinly veiled in reasoned discussion. Before long, though, their grievances strain the limits of polite conversation. Then there is the feeble acknowledgment by either sex of legitimate points made by the opposite side. Next comes finger-pointing, especially claims of bad faith on both sides of the gender divide. By then, nothing can stem the tide of ill feeling unleashed during such bruising encounters. The result is often a shrill combination of emotional catharsis and intellectual confusion. Much heat but little light escapes.

Long before the Clarence Thomas-Anita Hill debacle, which finally put some of the thorny issues between black men and women on the map, there was plenty of proof that many of us were thinking in harmful ways about gender. Many black men and women believe that placing questions of gender at the heart of black culture is an act of racial betrayal, a destructive diversion of attention away from race as the defining issue of black life. There's no denying that race has been, and remains, the Rubicon that black Americans must cross to arrive at healthy personal and social identities. The list of those who have argued that black people should subordinate race to, say, class or ethnicity in explaining our condition is long. And their arguments are largely unconvincing.

It's important to note that those sorts of arguments aren't the same as believing that factors besides race are equally important in determining the quality of life for black folk. I certainly believe that race remains a crucial part of the story of what's wrong with the treatment of black folk in America. I just don't think race is the complete story. There's too much evidence that being gay, or lesbian, or female, or working poor makes a big difference in shaping the role race plays in black people's lives. Only if we're seduced by stereotypes that distort black culture's beautiful complexity can we believe that we're all the same.

Unfortunately, differences within black life have been viewed by many leaders and intellectuals as signs of our surrender to the corrupting influence of the white world. If you buy this logic, it wasn't until white women borrowed black styles of protest in the 1960s, and attracted black women by dressing their anger in political speech, that the perils of feminism crept into black culture. Never mind the heroic precedents of Anna Julia Cooper or Ida B. Wells-Barnett in the nineteenth century, or Ella Baker and Fannie Lou Hamer at the height of the black freedom movement thirty years ago.

The same provincial perspective has proved true for issues of sexual identity as well. On this view, homosexuality is for white folk, a perversion they invented because they had too much leisure time gained in pillaging and plundering nonwhite peoples. Not only is being a "fag" or a "dyke" downright unnatural, such logic goes, but it misrepresents sexual traditions of black American society, and before that, African culture. To believe that, you've got to overlook the brilliant contributions and sexual habits of Alain Locke and James Baldwin, or Bessie Smith and Audre Lorde. The differences that these differences make within black culture are too important to simply be dismissed.

The truth is, Marcia, that for the most part black men have been unwilling to confront inequities between ourselves and the women in our lives, inequities that we deeply invest in and justify by all sorts of philosophical and rhetorical gyrations. I guess the bottom line is that as black men we have a hard time seeing ourselves as oppressors, as doing to black women what has wrongfully been done to us. So many of the black women in our lives understand our difficulties, overlook them at times, even, regrettably, make their peace with them, thereby inviting their own downfall.

Marcia, I realize that you know all of this. After all, you've made such choices in the past and endured the suffering that literally tagged along. The physical and psychic abuse you encountered in your first marriage has left its mark. It has sometimes made you reluctant to speak your mind, to clear your throat and let your vocal cords vibrate with justifiable rage. So many women are caught in this bind of being violated and voiceless. Thus these women's distress is compounded: first as they are abused, and time and again as they're rendered incapable of verbalizing their traumas. When abused women nurse their wounds in enforced silence, they end up reliving, not relieving their suffering.

Of course, you've never painted yourself as a helpless victim. I'm not slighting the spunk and stamina you must have marshalled to make it through those tough times. I know, too, from my own mistakes that relationships are a two-way street, that both parties do their share of fouling things up. But nothing justifies the sort of abuse you've described to me. How humiliating it must have been to get your husband and kids dressed to attend a Thanksgiving meal at your mother-in-law's home, only to be *told* at the last minute that you couldn't go because you were being punished. How terrible it must have been, too, to love as deeply and devotedly as you did—a habit you certainly haven't forsaken—only to be occasionally choked and punched.

I hope it's clear that I greatly admire your determination to free yourself from such demoralizing conditions. I have learned volumes about the word courage as I glimpse your face after some ancient wound elbows its way into your life and you refuse it the time of day, though your gentle tears at night confess that its wearying work has not entirely failed. But you get up again in the morning, with a rare defiance of pain that black women have learned to master, your triumphant smile declaring victory over bitterness and despair. Your example has taught me a great deal about transforming suffering into a song. Equally important, you have reminded me that we must see in our misfortunes the footprints of a complex destiny.

Unfortunately, most of us have failed to see the larger meanings of gender reflected in the intimate details of our often confused love lives. I suppose it's no surprise then that I wasn't even aware of heady debates about the seductions and complications of gender before I fell in love in Detroit in the mid-'70s. My first sexual experience, ironically, was with a girl who belonged to my church. I say ironically because it was in church that I learned that I wasn't supposed to have sex until I was married. I guess the tension from all that repressed libido got the best of us.

Although she was my age, Rhonda was already experienced when she became my second girlfriend. She led me into the vineyards of erotic pleasure when I was a few months shy of my eighteenth birthday. In my neighborhood, I was definitely a late bloomer. Rhonda had a pretty face and a sweet, winning personality. She also had a pair of shapely hips. I learned to long for their gentle undulations as she taught me how to make love. Thrust here, stroke there, now slower, now faster, oh yes, right there. I had often heard the fellas in my neighborhood talk about being "pussy-whipped," of being so enamored of the sexual pleasure girls offered that you'd do almost anything to keep getting it. I was now a proud member of a fraternity that I later learned included most heterosexual men.

To be sure, I had pangs of guilt. After all, I still sung in the choir with Rhonda, and we heard countless warnings from adults around the church against premarital sex. One mentor offered me a priceless pearl of wisdom that portrayed the perils of anonymous sex: "Any pussy not surrounded by a person has no high purpose." I still can't get that alliterative aphorism out of my mind, especially since it conjures visions of a disembodied sex organ. Nevertheless, our elders' warnings failed to stop us from "doing the nasty." Besides, I was deeply in love. Since Rhonda took birth control pills, we didn't use a condom. This was shortly before the dawning of the Age of AIDS. Such mistakes were not punished then, as they are now, with the possibility of imminent death.

For a couple of years before I had sex, I heeded the advice of another of my mentors by purchasing *Playboy* magazines to tame my hormonal rage. The implicit moral message in this advice was clear enough: masturbation is better than fornication, or as many church folk mispronounce it, "fornification." I worked out a routine of autoeroticism that was quite effective, and on one occasion, quite embarrassing.

Once, after speaking to Rhonda on the phone, I retreated to my room at Cranbrook, a prep school in the suburbs of Detroit I was then attending, to engage in my ritual of release. I was still a virgin. Just as I had begun, my roommate from the previous year burst through my door without knocking. The shocked expression on his face was greeted by my look of deep embarrassment. The next time I saw him, the incident never came up. In fact, we never discussed it at all. But our relationship changed. We never quite got back on an even keel. I suspected that he told the other guys in the dorm that I had a peculiar fetish for down feathers.

My relationship with Rhonda ended when she went off to college and began to have sex with another guy. She told me it was her "play brother." I guess Carol Stack's notion of "fictive kin" gave her no pause, proving that lust is thicker than fake blood ties. I was crushed. It was her second infidelity. (Her first was with another college chum named Mike. I couldn't entirely blame her; at least she stuck with the same name!) I'll never forget her mother lamenting to me, after my discovery of Rhonda's disloyalty just before we broke up, that she wished Rhonda could find a "real man." Her mother was a member of my church, too. Her words had weight, enough to deflate my ego for a long while.

Several months after I broke up with Rhonda, I fell in love with another woman from my church. By the time I met Terry, I had been kicked out of prep school, and I had just received my high school diploma from "night school." I was eighteen. She was twenty-six. Terry was an actress, dancer, and model. A native Detroiter, she had recently returned from New York where she had a bit part on Broadway, had appeared on the cover of *Ebony* magazine, and had snagged small roles in a few off-Broadway plays. Terry had an arousing sensuality that I found intoxicating. I hungered for her immediately. I fell under her spell through exotic stories she told me about life in the Big Apple, a city I longed to visit.

Because she was more mature than I Terry radiated an air of sexual freedom and confidence that sharply contrasted to my relative naivete. I can't say for certain what she saw in me. Perhaps it was my innocence, even my ignorance. Maybe it was my fawning curiosity about her life. Or my avid appreciation of the guts she had shown in toughing it out in a brutal, bustling city like New York. Mostly what brought us together were our endless conversations about what we planned to do with our lives, and the dreams we kept hidden until some understanding soul happened along who wouldn't laugh out loud at our ambitions. Terry, of course, wanted to become more successful as an actress and dancer. I wanted to be a preacher. We applauded each other's aspirations. It was only later that we would come to tear at each other's throats with a vengeance that only former lovers can know.

For the time being, though, we became fast friends. As we told everyone in our church, and as we said to each other, we were "companions in Christ." It was a classic case of evangelical evasion of fleshly desire. In the meantime, I lusted after Terry, spurred on by a publicity photo that featured her in a skimpy bikini. At church, one of the younger staff ministers looked past me and Terry's confession of platonic affection and glimpsed the erotic energy that boiled barely beneath the surface.

"Don't y'all be over there screwin'," he half-jokingly chided us.

His terse warning proved to be prophetic. Not long afterward, Terry and I made love. Soon she was pregnant. Terry was only the second woman I had made love to, and now I

was going to be the father of our child. I was still only eighteen. In many ways, I was a child myself. As I had heard older
black southerners say, "I was so green that if I had been
planted in the ground I would have grown!" Many folks
thought that at twenty-six, Terry should have known better,
should have used birth control or made me use a condom to
prevent a pregnancy. But I was so in love with her by then
that I thought we should get married. We did. It proved to be
a serious error.

For starters, we were poor. Although Terry worked as a
waitress, morning sickness soon forced her to give up her job.
The two full-time jobs I worked barely allowed us to make
ends meet. Although I eventually quit those two jobs when
Terry's uncle helped me land a job as a clerk at Chrysler, I
was fired in less than three months. I hadn't been allowed to
work long enough to receive insurance. The only explanation
my white boss gave me for my termination was: "The company needed to fire somebody. It had to be somebody's ass,
and I'd rather it be yours than mine." Period. Terry and I had
to apply for food stamps, and then AFDC. We also benefitted from the WIC program. I continued to hustle odd jobs
here and there, from painting houses to cutting grass and
shoveling snow, just to make a few dollars.

On top of our money miseries, we found it hard to get
decent housing. We lived first in Terry's apartment, but we
were literally evicted on Christmas Day. After securing
boarding in another ghetto apartment building, we witnessed
a man kicking down the door of our future next-door neighbor, whipping out a gun and filling the air with vengeful
obscenities. We stayed for one night. We finally found shel-
ter in a house that was, ironically, across the street from
where I had once lived as a child. This dead-end street was
mean, the site where my "play sister" had been strangled to
death several years before.

We lived upstairs in a crumbling two-family flat, except we
shared our portion with a grouchy old black man named Mr.

Everett. He had his own room, and Terry and I shared two small, deteriorating rooms. We all shared the same bathroom. We barely saw Mr. Everett, but when we did, it was often ugly. Once he pulled a gun on me in a dispute about his access to the bathroom. Most of the time, he just cursed and lived in his own cramped world. Few people came to visit him. He was a bitter, lonely black man.

Terry and I weren't alone, but our relationship had certainly soured. We couldn't afford a bed at the time we lived in the two-family flat so we plunked a mattress on the floor. It was on that mattress, two months after we'd been married, that she casually made a statement that shocked me, and revealed the roots of her festering resentment of our lives.

"I don't love you," she confessed. "I never loved you. I shouldn't have married you."

My mind began spinning as my heart sank beneath my feet. I was deeply in love with Terry. This wasn't how it was supposed to be. I imbibed those lessons at church about how Christians weren't supposed to get a divorce, how they should work things out, how they should fast and pray until they made it right, until God fixed it up. But Terry was having none of that.

"We should pray about it," I said. "Maybe things will get better if we just pray about it. We can work it out."

"But I don't love you," she insisted. "I mean, this just ain't gettin' it."

"But I love you," I said as I began to cry. "I really love you."

But it was no use. From that moment on, our relationship was doomed. Shotgun marriages, I later learned, often backfired.

Terry turned vicious. She compared me unfavorably to her former lovers, told me I'd never be anything, said that I was a loser. I sure felt like one. I had been a ghetto kid who had made good by winning a scholarship in 1976 to a prestigious prep school, with the help of my first girlfriend's father (who was a prominent federal judge) and because of my high test scores and grades. Still, I had to repeat the eleventh grade at Cranbrook. I went to school with the sons of ruling class elites. It was the first time I had shared classes with white kids.

At Cranbrook I faced for the first time unblinking snobbishness from some of my classmates. I also received encouragement and acceptance from other students, and from many of my teachers. Unfortunately, the racism directed at the ten black males who attended Cranbrook was so intense—this was right when Alex Haley's *Roots* was televised, and among many harsher incidents, a note appeared on my door, "Nigger go home"—that I felt terribly isolated and assaulted. I had failed miserably.

I had to return to the same Detroit ghetto that my admission to Cranbrook had been a ticket away from. It was humiliating. My bright promise now turned to dark foreboding. What would I do? Would I go to college in Detroit, now that Harvard, Princeton, or Yale seemed beyond my grasp? It was shortly after my return to Detroit that I met Terry and got her pregnant. At least I had her love, I believed. But now, even that cushion had turned to stone. My heart continually ached.

Terry and I split up a few months after our son, Michael, was born, but not before we learned to hate each other. She was the first woman who made me want to learn how to cuss, to damn her for her vicious, brutal behavior.

"Nigger, you ain't shit," she'd often holler at me when we fought. "You need to take yo' ass somewhere and learn how to be a man."

"I was tryin' to be a man when I married you," I'd shoot back. "I was tryin' to take care of my responsibility, and to show you that I loved you."

"Yeah, that would've worked if I loved yo' ass, but I don't."

"Bitch," I screamed, tears streaming down my cheek.

I felt mortified in using that epithet. I'd never said it before, ever in life. I was afraid that it held some power to nullify my standing as a good Christian, as an upright man.

"Muthafucka, please," Terry taunted me. "That's why I don't want to be married to yo' raggedy ass. You're pathetic. You're weak."

I could feel my blood boiling. More than that, I felt my dignity slipping away, felt my life pouring out drop by bitter drop. Even though I loved Terry, I now despised her, but not nearly as much as she despised me. Unfortunately, this emotion in one guise or another pretty much defined our relationship for many years to come. After our divorce, Terry denied me the right to visit Mike, especially after I had moved away from Detroit to attend Knoxville College in Tennessee.

"Unless you bring yo' ass to Detroit, you can't see him," she insisted. And she had the legal right to prevent me spending time with Mike. I hadn't fought her for custody of our son. I naively believed that she would allow me liberal visitation.

"Mike ain't visitin' you," she acidly asserted. "I got custody muthafucka. If you want to see him, bring yo' ass to Detroit. But he don't really need you no way."

That's the way it went for the most part until I sued her years later for visitation. I thought often during my struggles with Terry of how black men are viewed as heartless, sexual predators who father children without a moment's thought about their welfare. And here I was trying to see my son, to lavish him with love and affection, because I surely didn't have much money to give him as a twenty-one-year-old college freshman.

Even when I got Terry in court after several years of seeing Mike for less than two weeks out of every year, I felt that I was badly treated by the white judge. By then, I was a graduate student at Princeton University, working toward a Ph.D. in religion.

"Your honor, thank you for this opportunity to address the court," I began. "When I stood before a judge a few years ago as I was facing a divorce, I was a factory worker who had been on welfare, with no formal education. Today, I am a

Ph.D. student at Princeton University. Although I can't give my son a lot of material gifts right now, I can offer him the love I have for him deep in my heart. Since I have been systematically and unjustly denied access to my son for so many years, except from the time he was three until he turned five, I ask that you grant me summer visitation, and every other holiday as well."

I felt it was a reasonable request. The judge then turned to Terry and asked her what she thought was fair.

"I think he should get every other holiday, and a month in the summer," she smugly offered.

"Granted," the judge announced as he hammered his gavel.

I was shocked. Although my analysis of gender wasn't very sophisticated, I did have a grasp of the basic tenets of feminist thought. And I felt then, as I do now, that no serious, self-critical feminist would have thought that the judgment was fair. I didn't know enough of the complexities of feminist theory to prevent me from viewing this event as simply another episode of emasculation. I thought, here it goes again: a sister in cahoots with a white judge, who was probably incensed that I had pulled myself up from poverty and that I'd gone on to do well. I was probably a threat to his vision of who black men are and how they should behave. How naive I was to pipe off about my Princeton pedigree, thinking it would impress him! A black man just can't get justice in America, I bitterly concluded. It may very well have been sour grapes on my part. But a part of me still believes that important points of my analysis are true.

In Tennessee, my personal and professional relationships gave me a more complex understanding of how gender analysis and feminist theory operated, or failed to work, in practice. I began my freshman year at Knoxville College at an age when most students were graduating. If my age wasn't enough to make me stand out, I was also a licensed minister. I joined the Mt. Zion Baptist Church and shortly became an assistant minister to the pastor. I was invited to preach at various churches throughout the area.

It was during one of my ministerial visits to a local church in 1980 to address a Saturday morning Baptist Association meeting about Job and suffering that I met Brenda. She was a ravishing beauty whose short cropped afro formed a halo around her angelic face full of light brown freckles. Perhaps most striking to me were her beautiful feet, generously displayed in her brown, open toe, high-heel, strapless slides, commonly called "mules." As she glided through the church, I found the clip-clop sound her mules made as they rhythmically met her heels erotically charged. (My foot fetish came from sitting at the painted red toes of my first grade teacher Mrs. Jefferson as she read us stories. In my mind, that forged a connection between eros and learning. At least that's what I tell myself now.) I asked Brenda for her number, and asked her to accompany me to a church where I was to preach the next afternoon. I picked Brenda up on Sunday in my used, raggedy green chevrolet Impala (where I was once forced to spend a few nights when I had no housing).

After I preached, I took Brenda to dinner, and later, we went to her apartment. We talked from eight in the evening until six the next morning. We spoke about God and scriptures, about gospel music and preaching, about my sermon which she liked a great deal, and about everything two people who instantly connect with each other talk about. By dawn, we surrendered to the seduction of having similar intellectual and spiritual interests that were suffused by the rush of sexual passion. We made love that morning after we talked ourselves out of our clothes and into each other's arms.

Although Brenda looked my age, I discovered that she was eleven years older. I was twenty-two, she was thirty-three. That didn't stop us from falling in love almost immediately, or from moving in together a couple of months later. My father died in 1981, a few weeks after I'd met Brenda. Her tender care and deep compassion were crucial in helping me confront the mountain of grief that lay before me. Brenda spent almost all of the hours she wasn't at work as a recovery room nurse in a Knoxville hospital helping me sort out my feelings for my father. She heard me and held me as I cried. I recalled for her my great love for my father. I recounted my painful past with him as well.

When Mike was three, Terry's career enjoyed a slight revival. She allowed Mike to come to Tennessee to live with me and Brenda. I had pined to see Mike's chubby, dimpled cheeks and curly hair on a daily basis, and to have something concrete to contribute to his upbringing. He was a wonderful mix of charm and chutzpa, a bright, loquacious child who brought joy to me and Brenda. Already, he showed an independence of mind and spirit that have never left him. Mike and Brenda hit it off immediately, and she loved him as she would have loved her own flesh.

After nearly a year of living together, Brenda and I were married. A year later, Mike was whisked away from us, and until I gained legal visitation, we wouldn't see him regularly for several years. Brenda and I continued to support and love one another, though I detected a darkness in her spirit, a brooding sense of foreboding that kept her, and us, from completely letting go, from enjoying total happiness. It was not until years later that we would discover that she was clinically depressed. A large part of her depression drew, no doubt, from the physical and mental abuse she had suffered in two previous marriages and in a string of unhealthy relationships.

But her depression was also increased by the evolution of my vocation from preacher to graduate student to college professor. At each stage, the character of our relationship was affected by our rapidly changing identities. When Brenda met me, I was a student and a preacher with an eye on the pastorate. At the time, I had a job cleaning and degreasing heavy machinery at Robertshaw Controls factory. I pulled the 3 to 11 P.M. shift while I studied philosophy and religion during the day. Brenda was a licensed practical nurse. Over the next few years, though, I longed to become a serious scholar and intellectual, and Brenda longed to become a registered nurse.

Real tensions surfaced in our relationship as my contact with women deepened over the next few years. As a preacher and pastor, I was often called upon to counsel women, and to interact with the stunning variety of intelligent and beautiful "sisters" who comprise the black church world. Women would press admiring notes in my hands after sermons, or they'd make their way to greet me in public with warm, wet kisses. Understandably, Brenda got mad. Both of us knew of the long, concealed tradition of the clergy and its sometimes abusive sexual exploitation of women. I didn't want to participate in that tradition, but I didn't kid myself that I wasn't vulnerable to its lure.

I learned, however, that even when you think you're doing right by female members as a male pastor, there were plenty of surprises in store. A few months after Brenda and I started dating, I was called to pastor a small Baptist church outside of Knoxville. I quit the factory and assumed my first full-fledged pastorate while attending Carson-Newman College. (I had been interim-pastor of the Mother Love Baptist Church for a year while its pastor, who was also a university professor, took a year's sabbatical to study at Harvard.)

I announced in a church meeting that I had proudly avoided the three big sins associated with many ministers.

"I haven't gambled my money away," I boasted. "I haven't drunk liquor. And I haven't slept with any women in the church." After the meeting, a woman approached me with a troubled look on her face.

"I enjoyed your presentation Reverend Dyson," she began. "I'm glad you don't gamble and you don't drink. But what's wrong with the women in this church"?

Although there were many rocky spots, created both by my youth (I was twenty-three years old) and the church's

hard traditionalists, I managed to remain there a year before
being called to a larger church in East Tennessee. Brenda
quit her job in Knoxville, and we took up residence in the
parsonage owned by our new charge.

At my larger parish, I attempted to implement the liberating principles of feminist theory I was then reading outside
of the standard curriculum. Carson-Newman, where I had
transferred to get a strong grounding in philosophy, was a
white, conservative, Southern Baptist institution. They were
not known for their liberalism on race or gender. In fact,
right before I assumed my second pastorate, I had been
kicked out of Carson-Newman for a year because I had
skipped the chapel requirement, protesting the dearth of
black ministers they brought to campus.

"Well, based on the numbers of black students here, that's
fair representation," a chapel official informed me.

Despite my youth, I knew that I couldn't just rush in and
change church traditions that had been defined over several
decades. After consulting with Brenda, I took a gentle
approach. I worked hard at my second parish to lay the
groundwork for major changes in how women were treated.
I started with the Deacon Board, which represented the spiritual wisdom of the church. Like the pulpit, the Deacon
Board was held in high esteem. I knew it would have been
biting off too much to get the church to ordain a woman to
the ministry. But I figured that I might be able to get away
with ordaining women as deacons. Brenda agreed.

I preached sermons about God's liberating power, about
how God had freed black folk from white oppression. I tried
to link the theme of liberation from *racial* oppression to liberation from *gender* oppression. I often talked as well about
class issues, and at times, about the importance of confronting bias against gays and lesbians. I didn't want to give
my parish too much to digest, so I focused largely on gender.
I talked about gender in Bible study. I talked about gender in
church meetings. I talked about gender in members' homes.

Mostly I tried to show that it was hypocritical for black folk to talk about how God freed us from racism if we continued to discriminate against women.

Finally, I presented the idea of ordaining women to the Deacon Board, who function as the de facto power base in many Baptist churches. Quite naturally, there were some concerns, but for the most part, things went off without a hitch. I invited the deacons, most of whom were men above fifty years of age, to help me select three good female candidates. After three women were selected, I put them through an intense training session that lasted several weeks.

Just as the church was preparing a service of ordination for the women, the local Baptist Association got wind of our plans and went berserk. Several ministers called members of my Deacon Board and warned them of the dire consequences of having women in leadership positions. The ministers also planned to kick me out of the Baptist Association, and rallied several of my members, especially deacons, to oust me from the church.

One Sunday morning as Brenda and I came to church, the locks to the front doors had been changed. Finally someone with keys arrived, and we entered the church and went to my office. As I entered the sanctuary to preach, I knew that my ministerial colleagues and disgruntled members had met to plan my quick departure because I saw angry faces of members that had never been to church during my one-year tenure. Right there on the spot, they held a meeting after a hostile morning worship service and voted to release me. The Deacon Board cut our telephone off the next day, gave us a month to vacate the church parsonage, gave me a month's severance pay, and never looked back.

I was most hurt by the fact that so many women had opposed me. Even the women I had helped train to be deacons had remained silent. I hadn't yet understood the complexity of social theories which explain how people are seduced into cooperating with their own domination. I hadn't learned that often women are the biggest roadblocks to feminist beliefs because they, too, have become comfortable with the deals of convenience they've cut with patriarchal power. Other women fear that if the applecart of masculine domination is upset then the resulting chaos will undermine their security.

Still other women become terrified when beliefs they've learned to accept their entire lives are challenged as illegitimate. And the plain fact is that many of these women were married to the men who spun awkward but effective justifications for their continued superiority. That and a lot more was happening, and I didn't understand it all then. I think I'm clearer about these matters now, but then I was deeply hurt that the very women who stood to gain most in the long run had turned on me.

I think similar reactions stunned Anita Hill on a much larger scale. All she was trying to do was to tell her side of a story that had been kept quiet, first out of understandable self-interest, and later by a terrible combination of female fear and masculine dominance. But many black men *and* women blamed her for betraying the race, for spilling her guts before millions in what they saw as an attempt to do a "brother" in. The Hill–Thomas hearings forced many of us to think about how sex and power are linked, how the bedroom and business are often fatefully intertwined. How we are reared as boys and girls affects what we think as men and women. Our intimate relations have much more to do with large features of our social world than many of us are conscious of or willing to admit.

The stress and strain of the conflicts at church certainly burdened my relationship to Brenda. And her depression only seemed to deepen. One of its side effects was a considerable diminishment of sexual drive. But all of this is hindsight. At the time, I simply thought that she was rejecting me sexually, a pattern that continued over the next several years as we returned to Carson-Newman, and as I pursued a Ph.D. at Princeton, and taught at Hartford Seminary and Chicago Theological Seminary.

The lapse in intimacy became a psychic spur for me. It irritated old wounds inflicted by Terry's rejection, and over the ten-year course of my marriage to Brenda, I strayed sporadically. Plagued by a severe case of guilt each time I was unfaithful, I was hugely unsuccessful in my extramarital escapades. Many times, I couldn't sustain physical stimulation. At other times, after an initial foray, I simply refused to return for more. The women with whom I was unfaithful were no doubt greatly dissatisfied by my dismal performances. And I was profoundly ashamed.

Brenda and I eventually fell apart under the weight of my own responses to our troubles, and the depression that besieged her, a diagnosis which, tragically, was made only after I had decided I couldn't go on. I had grown up with Brenda, had loved her deeply, more deeply than I had loved any other woman. Her keen intelligence was continually undermined by a chronic lack of esteem deposited by the bitter experiences of her life and exacerbated by her depression. Her lack of healthy self-regard played a large part in her inability to believe that I could love her, should love her, that she was worthy to be loved. That, coupled with my own deep insecurities and pains from rejection, formed a fist to wreck our home. Our breakup after we had moved to Chicago was painful and protracted, a hurt beyond any hurt I've known.

Marcia, I vividly recall the moment I first saw you in all your splendid, breathtaking beauty. Your hair was classically coiffured in an elegant French roll. Your deep set eyes were smiling, your gorgeous caramel skin glowing. Your looks were at once sensual and regal. But most of all, I was taken by the perfect proportion of your lithe, exquisitely wrought frame, and your beautiful big behind. I wrote an amateurish poem celebrating your body. Part of it said:

Music in your body
is your body

Lilting waves of graceful movement
You choreograph space
with rare charm and ease

Your strong, beautiful legs are
sensuously compact and elegantly honed
from the hand of a God who signatured
an amen of approval in your smooth calves

And what of your portioned missive
of sensual delight?
That orbed oracle of sweet affection
speaking volumes of motioned seduction
resting charitably between
Your back and legs
(that I want to rest in my appreciative hands).

Marcia, as I think of how our relationship has grown over the last five years, I'm overwhelmed with gratitude. To be sure, I haven't figured out this age business completely, since you're eight years my senior, making it the third straight time I've married a woman older than I. But then, I didn't know that when I met you since you look younger than me. I'm just grateful for the care and compassion you show me. My past hurts, and the complex social world we have inherited—especially as we as a people, and a nation, attempt to get the gender politics right—prods me to salute you. I want to thank you for the imaginative and demanding ways you have loved me and encouraged me to grow. You haven't for a moment permitted me to dodge the painful contradictions between my proclamation of feminist principles and my inability or unwillingness, sometimes, to live up to them. But your patient love has made me a better human being.

Even though I make a living using words—writing them in private and speaking them in public—trying to express my love for you has made me painfully aware again of the limits of language in saying what is really in my heart. There is so much I want to say to you, so much I want you to feel. I confess my love for you. And I acknowledge how your love has changed my life and made it better! In the deepest, darkest hours of my suffering, I know your love has helped save me, showing me a better self and way of life than I had known or could imagine alone.

Your companionship and friendship have been the breath of a new beginning, a fresh start in the midst of failure and fear. You have loved me with tenderness and tenacity, showing me a loyalty so fierce that I can barely describe its effect on my being. You have renewed in me a confidence that what I am doing is good and that as a human being I am really alright. Neither money nor any of this life's material rewards can purchase such love and devotion. Even now tears of joy and appreciation crowd my eyes as I write to you.

I have learned, too, about spirituality from your life, though I am the professional and you a devout and classic amateur. Like Edison was the amateur who discovered the means of spreading light while the professionals remained in darkness. Your ease of turning to God, your seemingly effortless resolve to make God the center of your universe, simply embarrasses me.

But I have watched you, and I know that such devotion comes at a price: of having to hold on to God when nothing or no one else would or could save you. When the bare edges of desperation were the only room you had to stand on. When the comfort of God's lonely voice was the only suitable company for a soul worn from worry about children, house and job. When only an Almighty God could nourish in you hope for life beyond the hell of misunderstanding and confusion into which you had fallen. Your life is a prayer, each day a mounting stanza of pure devotion to a God who refuses to be distant from the center of earthly calamity. A God who gives joy and comfort as a reward to those who obey. Your obedience to God is light and lesson for me.

But most of all, I have learned to love myself more because of you. I remember when you first bathed me, and placed on my body the beautiful multicolored robe that symbolized for me life's gift and promise. Tears streamed down my cheeks, for I had never before been so moved by the expression of unhindered affection and care as you displayed. You scrubbed my body and massaged my pain. And as you looked into my eyes, you reminded me (but the truth is I learned it from you for the very first time), that I should never again become hostage to another's love for me, not even yours. I have never forgotten that encounter, a touchstone in my maturing as a man, and our relationship as a couple.

I guess by now it is obvious that you have been my closest friend. Often husbands and wives miss the real joy due to them by obscuring the fact that finally they are each other's bread and water. I say this from experience, because I know what it means to be each other's ball and chain. Of course, there is enormous risk in such intimacy. But the fruits of a deeply shared life far outweigh those risks. I shudder when I think of some of the things I have shared with you, and the vulnerability such sharing entails. But such thoughts are quickly disposed of when I think of the gentleness you have displayed about my doubts, the sensitivity you have shown in the face of my fears, and the unspeakable love you have shown in the face of my weaknesses and failures. And yet you remain my soul's closest companion, and for that I have eternal gratitude.

My love, there is so much more to say, and I will not try to say it all here. I will try to show it to you as we live and love together. As we continue our journey together, I will show you the love you have shown me, a love so large and understanding that it is able to vanquish the vilest opposition and to remain a source of comfort and strength in times of trouble. Yours is also a love which knows how to celebrate happiness and appreciate joy. In a real sense, I know we are already a team, the A and B of a complex puzzle of togetherness. Already, when people see one of us, they ask about the other, knowing we are never far apart from one another.

There is a comfort in knowing you are always near, that you want to remain close to me, as I want to remain in the precincts of your fragrant presence. I imagine when some people hear me declare in public that I could never achieve what I do without your love and support, they think it is sweet and thoughtful that I acknowledge you. But for me, it is perhaps the deepest truth of my life and career that you mean so much to me, that without your patience and persistence, I would be much less of a man. I am more complete with you, feeling the urgent press of your sweet flesh against my body, feeling the bright glow of your pride in who and what I am.

Recently, when you had abrasions on the corneas of both your bright eyes and you couldn't see for a few days, I had the privilege of taking care of you, of feeding you, dressing you, guiding you, of wiping your hands when they were dirty and brushing the crumbs away from your mouth, of combing your hair and kissing your forehead to let you know it would be alright. And then it struck me that this is what you do for me on a daily basis. You care for me and about me with such complete attention that my every need is met, my every desire fulfilled. Thank you for such unblushing commitment to me, such rare devotion. I pledge here and now to return such love and devotion with joy and passion.

Finally my sweet, let me tell you here, before this mighty tide of witnesses, that it is you that I want, you that I need, you that I adore, you that I desire, you that I dream of, you that I love. Even now, the mere scent of your perfume stirs in me deep urges of affection. Even now, the faint glint of your smile across the room reassures me of your undying devotion. Even now, the gentle caress of your diminutive hands reminds me of the ocean of care that backs every gesture of love you extend. Even now, in the midst of a sea of intelligent and beautiful women, you are *my* ebony fantasy enfleshed in caramel colors, a black beatitude of hope that brings blessing to my life. Even now, with all I have to choose from and with, I still choose you. And I hope, with all that you have to choose from and with, you still choose me.

Love, Your Husband,
Mike

Index